Guidelines to Minimize Downtime During Pipe Lining Operations

Awwa Research Foundation

About the Awwa Research Foundation

The Awwa Research Foundation (AwwaRF) is a member-supported, international, nonprofit organization that sponsors research to enable water utilities, public health agencies, and other professionals to provide safe and affordable drinking water to consumers.

The Foundation's mission is to advance the science of water to improve the quality of life. To achieve this mission, the Foundation sponsors studies on all aspects of drinking water, including supply and resources, treatment, monitoring and analysis, distribution, management, and health effects. Funding for research is provided primarily by subscription payments from approximately 1,000 utilities, consulting firms, and manufacturers in North America and abroad. Additional funding comes from collaborative partnerships with other national and international organizations, allowing for resources to be leveraged, expertise to be shared, and broad-based knowledge to be developed and disseminated. Government funding serves as a third source of research dollars.

From its headquarters in Denver, Colorado, the Foundation's staff directs and supports the efforts of more than 800 volunteers who serve on the board of trustees and various committees. These volunteers represent many facets of the water industry, and contribute their expertise to select and monitor research studies that benefit the entire drinking water community.

The results of research are disseminated through a number of channels, including reports, the Web site, conferences, and periodicals.

For subscribers, the Foundation serves as a cooperative program in which water suppliers unite to pool their resources. By applying Foundation research findings, these water suppliers can save substantial costs and stay on the leading edge of drinking water science and technology. Since its inception, AwwaRF has supplied the water community with more than $300 million in applied research.

More information about the Foundation and how to become a subscriber is available on the Web at **www.awwarf.org.**

Guidelines to Minimize Downtime During Pipe Lining Operations

Prepared by:
Thomas D. Rockaway
University of Louisville
Civil and Environmental Engineering, Louisville, Kentucky 40292
and
R. Timothy Ball
Infrastructure Renewal Services, LLC
12610 Old Henry Road, Middletown, Kentucky 40223

Sponsored by:
Awwa Research Foundation
6666 West Quincy Avenue, Denver, CO 80235-3098

Published by:

Awwa Research Foundation

American Water Works Association

IWA Publishing

DISCLAIMER

This study was funded by the Awwa Research Foundation (AwwaRF). AwwaRF assumes no responsibility for the content of the research study reported in this publication or for the opinions or statements of fact expressed in the report. The mention of trade names for commercial products does not represent or imply the approval or endorsement of AwwaRF. This report is presented solely for informational purposes.

CONTENTS

LIST OF TABLES

LIST OF FIGURES

FOREWORD

The Awwa Research Foundation is a nonprofit corporation that is dedicated to the implementation of a research effort to help utilities respond to regulatory requirements and traditional high-priority concerns of the industry. The research agenda is developed through a process of consultation with subscribers and drinking water professionals. Under the umbrella of a Strategic Research Plan, the Research Advisory Council prioritizes the suggested projects based upon current and future needs, applicability, and past work: the recommendations are forwarded to the Board of Trustees for final selection. The foundation also sponsors research projects through the unsolicited proposal process; the Collaborative Research, Research Applications, and Tailored Collaboration programs; and various joint research efforts with organizations such as the U.S. Environmental Protection Agency, the U.S. Bureau of Reclamation, and the Association of California Water Agencies.

This publication is a result of one of these sponsored studies, and it is hoped that its findings will be applied in communicating throughout the world. The following report serves not only as a means of communicating the results of the water industry's centralized research program but also as a tool to enlist the further support of the nonmember utilities and individuals.

Projects are managed closely from their inception to the final report by the foundation's staff and large cadre of volunteers who willingly contribute their time and expertise. The foundation serves a planning and management function and awards contracts to other institutions such as water utilities, universities, and engineering firms. The funding fro this research effort comes primarily from the Subscription Program, through which water utilities subscribe based on their annual billings. The program offers a cost-effective and fair method for funding research in the public interest.

A broad spectrum of water supply issues is addresses by the foundation's research agenda: resources, treatment and operations, distribution and storage, water quality and analysis, toxicology, economics, and management. The ultimate purpose of the coordinated effort is to assist water suppliers to provide the highest possible quality of water economically and reliably. The true benefits are realized when the results are implemented at the utility level. The foundation's trustees are pleased to offer this publication as a contribution toward that end.

Walter J. Bishop
Chair, Board of Trustees
Awwa Research Foundation

Robert C. Renner, P.E.
Executive Director
Awwa Research Foundation

ACKNOWLEDGMENTS

The authors of this report thank the Awwa Research Foundation (AwwaRF) for its support of this study, and the AwwaRF project managers Stephanie Passarelli and Jian Zhang. The authors would also like to acknowledge the input and input and guidance throughout the course of the project from the members of the AwwaRF Project Advisory Committee:

- David Hughes (American Water Company)
- Kenneth Morgan (Private Consultant)
- Mark Warriner (Metro Water Services)

A special thanks for the dedicated efforts of our participating Utilities for the hours of data tracking, gathering, and review. In particular we would like to thank those involved with the sponsor utilities, Pat Arnette with Greater Cincinnati Water Works, Keith Coombs with Louisville Water Company, and Todd Fuller at Northern Kentucky Water District, without whose dedication, support, and input we would not have been able to successfully complete the study.

In addition to the above listed, many other utilities provided data, participated in a workshop and/or participated in the disinfection study. We greatly appreciate the assistance and contributions from James Matt with Anne Arundel County, Maryland Department of Public Works; Michael Henderson and Nwamaka Dallah from the City of Cleveland Division of Water; John Veleradsky from the Metropolitan Utilities District in Omaha, Nebraska; Allen Brown of Charlotte-Mecklenburg Utilities, Charlotte, North Carolina; Nicholas Nicolaides on behalf of the Bureau of Water and Wastewater, Baltimore, Maryland; Bret Russell from the Louisville Water Company, Louisville, Kentucky, Jon Sullivan with Metro Water Services, Nashville, Tennessee; Kenny Treadway of the City of Greensboro Water Department; John Allen from the City of Fayetteville Public Works Commission; Austin O'Blenis from the City of Moncton, NB; and Brent McGovern of the City of Saint John, NB.

We would also like to acknowledge the assistance provided by the University of Louisville Research Assistants with a particular gratitude to Jason Bush for his insightfulness with regard to data analysis and display.

EXECUTIVE SUMMARY

RESEARCH OBJECTIVES

Repair and maintenance of existing distribution lines has been an ongoing concern for utilities since the first network was placed into service. Infrastructure rehabilitation needs continue to increase and utilities have looked to a variety of methods to keep pace. One of the more common renewal methods employed by utilities is pipe lining. While pipe lining activities are effective and generally less expensive then complete replacement, they still represent considerable time and expense. Thus, many utilities have developed their own protocols for pipe lining operations that have helped them to improve performance and reduce costs for their specific applications.

The purpose of this research initiative is to provide guidance on the best practices to minimize the downtime associated with pipe lining operations. In this context, "downtime" refers to not only the amount of time customers are without water service, but also the amount of time customers must rely on temporary service, endure restricted service, or are served by less secure networks. The research work critically reviewed current pipe lining operations and a statistical analysis was used to identify efficient operations and procedures.

Based upon findings of the study, this report presents "checklists" designed to optimize the specific pipe lining operations. The checklist is supplemented with a "commentary" that discusses each checklist item and presents reasoning and alternative procedures where practical. By using the checklist and commentary, utilities should be able to complete projects without critical omissions and incorporate best management practices as appropriate for their specific applications. Additionally, the checklist and commentary has been incorporated into a project scheduler template for use with Microsoft Project. The intent is that utilities can then adjust the base template such that it is specific to their application.

This research study benefits utilities interested in pipeline rehabilitation by providing best management practice information and scheduling tools that optimize the time associated with pipe lining operations. Opportunities to improve the efficiency of rehabilitation tasks are presented to help utilities compress the construction cycle, avoid common pit falls and ultimately minimize the "downtime" associated with pipeline rehabilitation.

APPROACH

To identify efficient pipe lining operations, both published literature and utility resources were utilized. To assess current "state of practice", published literature, case studies and manufacturer brochures were reviewed. However, the technical data to assess lining operations was primarily provided by nine utility participants actively engaged in the lining process. The participating utilities maintained detailed records of their pipe lining operations with respect to schedule, task item, manpower requirements, and production rates. Specific project records were then "mined" to extract data such as duration, footage, problems, location, methods and techniques. In this manner, statistical comparisons between projects could be objectively evaluated.

Once industry trends were identified, the utility data was used to compare specific utility operations to the average values. The focus of the analysis has been to identify the underlying reasons utilities select specific design methods or techniques in response to risk, time, and monetary constraints. The "checklist", commentary and scheduler tools were developed based on the collected data and discussions during a facilitated workshop. The intent is for these tools to be used as a guide to help project managers properly orchestrate rehabilitation projects from design stage through construction.

CONCLUSIONS

Maintaining water distribution system integrity and increasing customer satisfaction are two critical components of water system operations. However, during maintenance operations, these two goals are often at odds. Maintenance operations typically require a service disruption, temporary service network and may include a significant amount of equipment or personnel in front of a customer's house. While these pipe renewal operations are necessary to provide reliable long-term service, customers must endure temporary service disruptions, construction equipment and construction noise.

The data collected during this project suggest that a number of best management practices have the potential to reduce both the length of time necessary for pipe lining operations and the length of time temporary pipeline is in service. Each recommended management practice would require a shift in what is currently considered a typical construction practice and acceptance of the associated risk. However, by combining a number of the recommended improvements to the cleaning and lining processes, the possibility exists to eliminate the need for temporary pipeline from the scope of traditional lining projects.

RECOMMENDATIONS

The following are recommended changes in the current construction processes that have the potential, if properly implemented, to greatly reduce the number of days most customers are subjected to maintenance activities and the temporary pipeline.

- *Improve Task Sequencing* – Consider each city block, or pipe segment, as a separate project with a reduced footprint. Each task should be schedule to follow as quickly as possible the preceding task so as to reduce delay and lag time. This reduced footprint construction sequencing does require an added level of attention to project planning and a greater level of scheduling manpower, however, it also has the potential to significantly reduce the amount of time customers are exposed to construction activities and placed on temporary service.
- *Change Lining Material* – The selection of appropriate lining method or material can significantly effect the duration of a pipe lining project. Both cement mortar and epoxy materials require approximately a 12-hour cure time prior to reintroducing water into the pipe. The application of a rapid set material, such as a polyurethane (PU), would allow for the water to be reintroduced into the pipe 1-hour after lining has been completed.
- *Change Temporary Pipe Material* – The data collected indicated that there is a substantial variation in installation rates for the three temporary pipeline materials commonly used. Of the products assessed, nitrile rubber was documented to require the least amount of manpower to install at a rate of 3.3-man hours/1,000 linear foot.

- *Rapid Chlorination* – The slug disinfection method (AWWA C651-99, Sec. 4.4.4) specifies a 100 mg/l initial concentration and a 3-hour contact time. Employing this disinfection method could significantly reduce downtime during lining operations by eliminating the 24-hour contact time associated with the tablet and continuous feed methods. The slug disinfection, if acceptable to local health officials and State Regulators, could be applied to both the newly lined pipe and the temporary water supply
- *Pressurized Hose Reel* – A pressurized hose reel that could be transported from block to block would eliminate the need for dewatering and chlorination at every reinstallation of temporary pipe. This would allow the temporary pipe to be disconnected from customer services and rolled onto the reel and transported to another nearby location without being depressurized. The savings for such a system would include some manpower that is typically required to install segmented temporary pipeline and the manpower required to chlorinate, flush, collect and incubate water samples.
- *Treat Lining Segments Similarly to Main Break Repairs* - Utilities that apply main break disinfection procedures to newly lined pipe can obtain a substantial time and costs savings. The main break disinfection procedure would eliminate the 24-hour time lag for both the chlorine contact time and the results of the bacteriological tests. Further analysis and demonstrated procedure effectiveness will be required by State Regulators before these processes are widely accepted and fully implemented. However, preliminary studies completed by six utilities, indicates that in most instances flushing alone will provide satisfactory disinfection for newly lined pipe sections (Appendix).

CHAPTER 1
INTRODUCTION AND BACKGROUND

PURPOSE

The purpose of this research initiative is to provide guidance on the best practices to minimize the downtime associated with pipe lining operations. In this context, "downtime" refers to not only the amount of time customers are without water service, but also the amount of time customers must rely on temporary service, endure restricted service or are exposed to less secure networks. The pipe lining guidelines presented herein are based on a critical review of the current pipe lining methods, materials, disinfection, and appurtenance renewal procedures implemented by participating water utilities. A statistical analysis of man-hour requirements and project durations are used to identify efficient operations and procedures.

Based upon findings of the study, this report presents "checklists" designed to optimize the specific pipe lining operations. The checklist is supplemented with a "commentary" that discusses each checklist item and presents reasoning and alternative procedures where practical. By using the checklist and commentary, utilities should be able to complete projects without critical omissions and incorporate best management practices as appropriate for their specific applications. Additionally, the checklist and commentary has been incorporated into a project scheduler template for use with Microsoft Project. The intent is that utilities can then adjust the base template such that it is specific to their application.

This research study benefits utilities interested in pipeline rehabilitation through best management practice information and scheduling tools designed to optimize the time associated with pipe lining operations. Opportunities to improve the efficiency of various stages of each rehabilitation method are presented and thus should help utilities compress the construction cycle, avoid common pit falls and thus minimizing the "downtime" associated with pipeline rehabilitation.

OVERVIEW

Repair and maintenance of existing distribution lines has been an ongoing concern for utilities since the first network was placed into service. Infrastructure rehabilitation needs continue to increase and utilities have looked to a variety of methods to keep pace. One of the more common renewal methods employed by utilities is pipe lining. While pipe lining activities are effective and generally less expensive then complete replacement, they still represent considerable time and expense. Thus, many utilities have developed their own protocols for pipe lining operations that have helped them to improve performance and reduce costs for their specific applications.

To identify efficient pipe lining operations, both published literature and internal utility resources were reviewed. Published literature provided many case studies and manufacturer brochures for evaluation and the references identified are presented in the report Appendix. However, upon analysis of the published literature, it was determined that most of it did not provide the level of detail necessary.

To assess current "state of practice", a survey was developed and distributed to select AwwaRF members and other key utilities. An invitation to complete the survey was distributed via email and a web based survey questionnaire recorded data. The survey questions were designed to identify general trends in the lining practice and to identify utilities actively involved in lining operations. Utilities that were actively lining were then asked to maintain detailed records of their pipe lining operations for further detailed review. Specific project records were then "mined" to extract data such as duration, footage, problems, location, methods and techniques. In this manner, statistical comparisons between projects could be objectively evaluated.

Once industry trends were identified, the utility data was used to compare specific utility operations to the average values. The focus of the analysis has been to identify the underlying reasons utilities select specific design methods or techniques in response to risk, time, and monetary constraints. In this manner, specific utility procedures were identified as efficient, necessary, and/or superfluous.

After the detailed analysis of the utility case studies, tools were developed to provide a "roadmap" for the experienced and the inexperienced Project Manager to execute a pipe lining project in an efficient manner. The tools consisted of a high level project "checklist", commentary and scheduler. The intent is for these tools to be used as a guide to help Project Managers properly orchestrate rehabilitation projects from design stage through construction. Proper sequencing is critical to overall project success as each task, if not properly addressed, has the potential to derail the momentum of the project, causing costly delays and erosion of customer confidence.

The results of this research project are presented in this guidance document for pipe lining operations. This document presents the efficient pipe lining methods, procedures and techniques as well as discussing efficient temporary service design. The goal of this document is to help facilitate an efficient and successful pipe lining operation.

LIMITATIONS

Much of this report focuses on cement mortar and epoxy lining rehabilitation techniques. While other pipe renewal techniques are available, these methods were not actively utilized by the utilities agreeing to provide information. It could be argued that some issues, especially those related to temporary services would be applicable for other rehabilitation technologies, such as bursting, sliplining, etc. However without the necessary performance data these rehabilitation methods could not be included in the analysis. The information presented in this report was obtained from published sources, participating utilities, and the researchers' experience.

PIPE REHABILITATION METHODS

Overview

Prior to 1940, most United States water utilities routinely installed unlined cast iron pipes throughout their distribution systems. These materials are still commonly in service; however, without internal protection they are susceptible to internal corrosion (tubercles). As the corrosion progresses, system performance degrades through observable reductions in flow capacity and reduced water quality. Cleaning and lining these problematic water mains has proven to be a cost effective way to restore high quality service and protect the interior wall of the pipe. A variety of cleaning and lining techniques are available to utilities for their rehabilitation efforts. Listed below is a summary of common pipe rehabilitation methods including cement mortar, epoxy resin, polyurethane, cured in place pipe, slip lining and pipe bursting.

Rehabilitation Techniques

A variety of pipe rehabilitation methods have been developed over the years and are generally separated into non-structural, semi-structural, or structural categories. All categories have very specific applications and characteristics that may make their use in a particular setting more or less efficient depending on the field conditions. Some factors which may affect the selection of an appropriate rehabilitation technology include: physical condition of the existing pipe, length of time it is necessary to maintain an alternate supply of potable water, surface and soil conditions, traffic, access, and number of service taps or connections. Most of these lining techniques, as well as decision guides to help select appropriate rehabilitation methods are available in AwwaRF publications such as:

- Assessment of Existing and Developing Water Main Rehab Practices, project #314
- Demonstration of an Innovative Water Main Rehab Technique: Epoxy Lining, project #808
- Demonstration of an Innovative Water Main Renewal Technique, project #255
- Guidance for Management of Distribution Operation and Maintenance, project #457
- Guidance Manual for Maintaining Distribution System Water Quality, project #357
- New Techniques for Precisely Locating Buried Infrastructure, project #2524
- Prioritizing Water Main replacement and Rehabilitation, project #459
- Quantifying Future Rehabilitation and Replacement Need of water Mains, project #265
- Rehabilitation of Water Mains: Manual of Water Supply Practices, M-28

Non-structural Pipe Lining

A non-structural pipe lining rehabilitation method involves applying a new surface to the interior of existing mains. The new lining does not provide an increase in pipe strength and relies on the integrity of the existing pipe for structural support. While the new lining will reduce the pipe interior diameter, increases in the Hazen-Williams C-factor can generally counteract potential reductions in carrying capacity. As stated by (Methods, et al 2003), non-structural pipe lining is generally considered a viable pipe line rehabilitation technique when:

- The pipe is structurally sound
- Future demands are not expected to be significantly greater than the demands for which the system was designed
- Loss in carrying capacity due to pipe tuberculation, scaling, or other deposits is significant (Methods, et al 2003)
- Water quality (red water) from interior corrosion is a principal problem.

A variety of materials and application methods are available to apply a non-structural lining on the interior of an existing main. Most non-structural linings utilize a spray applied:

- Cement Mortar
- Epoxy Resin
- Polyurethane

Within the United States, however, the most commonly used non-structural lining method is spray applied cement mortar. With this method, the cement mortar is thoroughly mixed and delivered to the lining machine in one of two ways; it is either pumped to the lining machine via a reinforced hose or it is carried by buggy to the machine. The pumped method employs a winch to draw the lining machine through the pipeline while the other is generally a self-propelled electric machine. In both scenarios the spray head is pulled through the pipe and cement mortar is "cast" centrifugally to the pipe walls. Either a cone shaped or rotating radial trowel closely follows the spray head to smooth out irregularities and to provide a uniform coating. Cement mortar can be applied to a wide array of pipe diameters (generally between 4 and 144 inches) and the speed of the spray head travel is closely regulated to provide the necessary cement mortar thickness. Once applied, the cement mortar lining is protected and allowed to cure for 12- to 24-hours.

The cement mortar lining method has proved to be particularly well suited to rehabilitation applications within the United States. The slightly higher pH of the green concrete helps reduce the likelihood of contamination during installation and the relatively minor long-term alkaline environment is compatible with most hard water systems. The cement mortar lining, however, is relatively thick (3/16-inch to 3/4-inch thick) and thus may plug laterals that are less than 2-inches in diameter if they are not protected during or cleared with air or water after troweling. Additionally, run lengths are generally limited to about 700 feet using the pump delivery system due to distance that mortar must be pumped to the spray head and turns or other obstructions. Under ideal conditions, however, cement mortar runs in excess of 1,500 feet have been performed with the mechanical mortar delivery method.

An alternative to cement mortar lining commonly used in the United States is spray applied epoxy. Spray applied epoxy resin was first used in England in the 1970s and subsequently used in the United States and Canada after about 1990 (American Water Works Service Co., 2002). Similar to the cement mortar method, the epoxy is cast to the pipe walls through a centrifugal spray head. The epoxy resin and hardener are supplied to the spray head in separate tubes and mixed near the nozzle just moments before spray application. The advance rate of the spray head is controlled by the winch speed and thus regulates the desired thickness. In general, epoxy coating thickness ranges from about 1 to 3 mm on the pipe walls. Once applied, the epoxy lining is protected and allowed to cure for 16-hours.

Due to the relatively thin cross sectional area, epoxy resin is often preferred over cement mortar when lining small diameter pipes or when flow volume must be preserved. This thin coating also does not block service or lateral connections and thus reduces the need for subsequent work after the lining process. However, uncured epoxy linings are susceptible to water-induced degradation and subsequently construction operations must ensure relatively dry pipe conditions prior to application of the resin and hardener. As an alternative to mechanically scrapping corrosion from the pipe wall and spray applying the epoxy coating some contractors may elect to use compressed air to both clean the pipe and to apply the epoxy material. For the cleaning, garnet is mixed with the compressed air and circulated through the pipe. The vortex action scours the interior pipe wall creating a smooth clean surface. After cleaning, compressed air is used to evenly distribute epoxy materials that are initially poured into the pipe.

Polyurethane is a relatively new spray application that has been adopted in England. Its application is almost identical to that of epoxy resin except that it is less susceptible to water intrusion due to its much faster cure time. In general, the most rapidly curing spray applied polyurethane is dry to the touch in less than 2 minutes and can be returned to water main service within an hour under ideal conditions. The material supply line to the spray head has also been modified to include a "heater hose". Elevating the temperatures of the polyurethane materials in the supply line helps to maintain the desired viscosity and thus increase run lengths and improve mixing.

Semi-structural Pipe Rehabilitation

Semi-structural pipe rehabilitation methods are used primarily when the host pipe is intact, but beginning to exhibit distress through pinhole leaks, joint leaks, or other minor failures. In general, semi-structural techniques involve creating a new, less rigid pipe within the existing pipe. This new less rigid pipe can either be pulled into the old pipe as a contiguous unit, or formed in place using the old pipe as a mold. The semi-structural lining may bridge small discontinuities or gaps in the host pipe, however, it is not intended to provide primary structural support. While semi-structural lining methods can improve the performance of existing pipe, the slightly reduced pipe diameter may adversely affect available capacity if not compensated by an increased C-factor of the lining material. Common methods of semi structural pipe lining include close fit (thin) pipe and thin-wall HDPE, Cured in Place Pipe (CIPP) or Woven Hose Lining.

The close fit (thin) and thin-wall HDPE semi structural lining materials are advantageous where access is limited and the host pipe is in reasonably good condition. These materials are generally pulled into an existing pipe as a contiguous unit thus providing a very clean uninterrupted surface. Often, outside diameter of these materials will be compressed slightly just prior to insertion and then expanded in place to form a very tight seal to the host pipe. These materials, however, are somewhat limited in their general application as they:

- significantly reduce cross sectional area,
- require a significant amount of construction room to lay out materials,
- limited to relatively straight installations,
- require significant additional work to reestablish connections and laterals, and
- may be damaged by sharp objects during installation.

The Cured in Place Pipe (CIPP) or Woven Hose Lining methods also create a secondary pipe inside the host pipe, however, these methods use the host pipe as a mold and use resin and a cloth mesh to provide structural rigidity. With these methods, a resin wetted "tube" is placed inside the host pipe and pushed to the existing pipe walls with vapor pressure. The pressure is maintained while the resin cures thus creating a bond between the host pipe and newly formed interior pipe. Service lateral connections are then robotically drilled from the inside or reestablished by spot excavations. The CIPP method requires significant construction expertise and is generally limited to larger diameter pipes or small diameter pipes with limited service lateral connections. Effective rehabilitation lengths, however, are only limited by the length of available tubes and ability to place the wetted material. These linings should be thoroughly researched, as they may generate future issues when attempting to make repairs or connections long after rehabilitation operations are complete.

Structural Pipe Rehabilitation

Structural pipe rehabilitation methods are generally applicable when the existing host pipe can no longer be relied upon to provide sufficient structural rigidity to maintain pipe integrity. This often occurs when the host pipe is suffering from severe external or internal corrosion or endanger of catastrophic failure. When structural alternatives are used to rehabilitate an existing pipe, the new pipe must have internal burst strength equal or greater than the original host pipe and may not rely on the existing host pipe for any structural support. A variety of methods are available to provide structural linings, and the appropriate method selection is dependent on specific field conditions. The commonly available structural pipe lining methods include:

- Slip lining
- Pipe bursting
- CIPP

Slip lining and pipe bursting involve pulling a contiguous pipe into an existing main. For slip lining, the outside diameter of the new pipe is reduced by about 10 percent (if necessary) and pulled into position. This method differs from the semi-structural slip lining method only in that a thicker or stronger pipe material is used. Pipe bursting methods are used when the new pipe has an equal or larger diameter than the existing main. A mandrel or cutting head is placed in front of the new pipe and it splits the existing pipe as it is pulled or jacked into place. Materials used during slip lining and pipe bursting of existing mains include: Ductile Iron, HDPE and PVC.

Structural CIPP or Woven Hose lining are similar to the products used in semi-structural application except that additional rigidity is applied to the system. The additional rigidity may be provided by substituting the woven tube with a fiberglass or carbon fiber mesh system. Multiple layers of mesh and different mesh materials can be used to tailor material properties to specific applications.

Cleaning Techniques

Prior to placement of any tight-fitting or spray applied pipe lining material, existing mains must be cleared of corrosion and debris. A variety of cleaning methods have been presented in other AwwaRF reports, and each has specific applications where they are more effective. Possible cleaning methods include flushing/jetting, scouring, chemical cleaning and mechanical scraping.

Flushing alone generally will not sufficiently clear a main for pipe lining operations. Flow velocities of 1.0 to 2.0 meters per second will remove silt and debris, but they will not dislodge corrosion or tubercules that commonly form in unlined cast iron pipe. Slightly better success has been observed when water pressures are increased and water jets are aimed at the corrosion features.

A variety of scouring methods have been employed to clean pipes. With a rapid flow of water passing through the pipe, sand, balls or other abrasive is mixed to into the stream to scour the inside of the pipe. An extension of this method is to repeatedly pass coarse swabs or pigs down the line until they abrasively remove all obstructions. Air scour is a less frequently used cleaning technique where high velocity, low pressure air is used to remove "sediment, biofilm and soft mineral scale" (Grob, 2004). Slugs of water are mixed with the airflow to create a vortex that scours the pipe walls. The process is continued until water exiting the pipe section is clear. While air scouring is reported to have a lower cost than other cleaning methods (Grob, 2004), hardened tubercules or corrosion features may not be adequately removed.

Mechanical scraping is the cleaning method most frequently employed prior to pipe lining activities. Mechanical scraping can be either in the form of a metal scraper disk that is pulled back and forth through the pipe and physically removes the debris. Alternatively metal scraping can be accomplished by a rack-feed bore machine that hydraulically extends a spinning rod with head flails down the pipe.

The most common cleaning techniques are covered in AwwaRF Manual 28: Rehabilitation of Water Mains – Manual 28, 2nd Ed., AwwaRF 2001. However, regardless of the cleaning method selected, the goal is to provide a clear, obstruction free pipe suitable for subsequent pipe lining operations. The effectiveness of these methods and cost varies considerably with pipe conditions and final objectives.

TEMPORARY WATER SUPPLY

Application

During planned service interruptions lasting for more than several hours, it is a common practice in the United States to install temporary piping to provide virtually continuous service to affected customers. The temporary pipe system is sized to meet the hydraulic requirements of the area and may include provisions for temporary fire hydrants as well as customer service. Typically, temporary supply lines are installed above ground and residences are connected to the system with flexible hose at outside spigots or in the meter pit.

Once temporary water lines are deemed necessary, project time, costs, and conflicts generally increase significantly. One of the main reasons temporary pipe is time consuming and expensive is that the temporary network must be disinfected and the water must pass bacteriological testing before customers can be connected. The temporary pipe installation and approval process adds to the project timeline considerably. The temporary lines can also be viewed as a public nuisance as they create a trip hazard for pedestrians, hindrance for vehicular traffic, and can be easily broken or vandalized. As such, temporary pipes are often the source of customer complaints during pipe lining operations and represent substantial risk to the utility.

Many other industrialized countries, including the United Kingdom, choose a less extreme alternative to temporary piping, depending on the length of the outage. These alternatives include:

- providing no alternative supply
- bottled water supply
- trucked-in water.

The decision to select one alternative over the others requires careful analysis of a variety of local factors such as site conditions, hydraulic or fire flow requirements, expected downtime, and the nature of customers who may rely on potable water for their daily operations, such as hospitals, schools, and industrially customers.

Temporary Pipe Installation

Installing temporary piping measurably increases project time, costs, liability, and scheduling conflicts. The principal factors that contribute to temporary pipe problems and expense include material selection, breakage, placement and duration the temporary pipe is in service. To reduce the risk and liabilities associated with temporary piping, the following should be considered:

- Install the temporary line where it is least obtrusive to the general public
- Install the temporary line where it is least likely to be damaged by vehicular traffic
- Install large enough lines to adequately serve the affected customers
- Install a water tight system to preserve drinking water quality
- Use a durable material that can easily be reused
- Use a material that is easily disassembled and reinstated
- Use a highly visible material

Temporary water supply mains are usually connected to fire hydrants just beyond the working area to maintain water service to the utility's customers. Connections at the hydrant may require the installation of meters and backflow prevention devices. Supply to individual customer services is typically made using an FDA or NSF approved 1- or 3/4-inch hose that runs from the temporary water main to either the meter setter or outside hose bib connection.

A variety of methods exist to insure safe passage over the temporary waterlines for vehicular and pedestrian traffic. Typically the pipes are buried flush with the surface in busy thoroughfares and commercial driveways. Prefabricated ramps (rubber, steel, or lumber) or temporary asphalt are usually placed at residential driveways where temporary pipes cross to protect these temporary mains from traffic and to provide safe vehicular access. Wheelchair access can be achieved by either burying the pipeline or constructing an ADA compliant ramp over the pipe.

The temporary piping must be disinfected and laboratory tested to meet water quality standards before placing it into service. This requirement to disinfect the temporary piping network and test for bacteria contributes to manpower demands and customer inconvenience, however, it does not increase the time the customer is supplied by temporary service .

Once installation and testing is completed, the temporary piping network provides continued water service throughout the project. Connecting the customer's water service line to the temporary main only interrupts the water service for a few minutes. After the rehabilitation of a section of the existing pipeline is finished, service lines are reconnected to the rehabilitated pipe, and the temporary water mains are removed and the landscape restoration process begins.

Governing Regulations

Water Pressure/Volume

In order to meet AWWA standards for drinking water systems, the pressure in the temporary water main should remain above 20 psi at all times. Beyond this minimum, the lines should supply the same pressure and volume of water to which the customers are accustomed. Reduced pressure or volume will typically generate customer complaints and result in reworking the system to adequately supply all existing services. Where the temporary supply is from distribution systems with elevated pressure, pressure reducing devices at temporary pipe feed points may be necessary to reduce the potential for temporary pipeline ruptures and internal piping damage.

Fire Protection

Utility requirements vary widely with regard to temporary supply system design for fire flow. Some municipalities may not require the inclusion of fire protection when developing the scheme for the temporary lines. However, other utilities can require full fire flow and a temporary hydrant installation for every permanent hydrant taken out of service. Generally local Fire Districts have a minimum allowable distance between hydrants or a minimum distance any structure can be from an operable hydrant. Proper construction sequencing and strategic preliminary hydrant installations can help reduce the requirements for elaborate temporary networks designed specifically to meet fire flow requirements. Additionally, advanced notification to and working closely with critical customers and local Fire Districts can reduce the need for temporary hydrants.

Water Quality

Most state regulations dictate that the temporary pipeline should be disinfected, flushed, and laboratory approved similarly to any newly installed permanent pipeline. Chlorine is the generally accepted disinfectant and AWWA allows for a variety of disinfection techniques and chlorine contact times per AWWA Standard C651-99. Additionally, as customers expect water supply to be provided within a reasonable temperature range, the lines may need to be continuously flushed during hot weather. Conversely, cold weather may require continuous flushing to prevent freezing. Some utilities randomly sample the temporary lines in much the same way they sample their distribution system to ensure water quality standards are met.

CUSTOMER RELATIONS

Historically, water providers within the United States and Canada believe that their customers expect uninterrupted service. For this reason a huge emphasis has been placed on installing and maintaining a temporary supply of potable water during routine maintenance operations. In the United Kingdom and other European communities, however, more of the utility's energy is devoted to educating customers on the necessity and benefits of rehabilitating water infrastructure as opposed to providing temporary service. For this reason, very little temporary pipeline is installed in these countries and customers are more tolerant of outages related to rehabilitation projects.

In the early stages of rehabilitating small diameter pipelines in the United States (i.e. the 1970s), it would not have been uncommon to give little or no warning to residents prior to beginning a project in their neighborhood. Gradually some utilities realized the benefit of interacting with customers before the start of construction. Thus, for current rehabilitation projects, it is not unusual to involve customers early in the rehabilitation process. Most utilities today send some type of notification through the mail a week or two before the contractor arrives on site. Additionally, many utilities will hold public meetings, advertise in neighborhood newspapers, and post notices in various public venues.

In the United Kingdom, utilities have taken customer relations a step further and actively educated their customers on the pipe rehabilitation process. Key points to be conveyed to the public include explaining why the process is vital to the utility's ability to supply the highest quality water into the future, what is to be expected with regard to the construction process, and the duration of expected water supply outage. The focus of the utilities' work then shifts to minimizing the construction time through planning and adaptation of new techniques.

EMERGING TRENDS

A variety of newer products are being adapted or improved for the water pipe rehabilitation industry. Among emerging techniques that are more commonly being incorporated to water mains are pipe bursting, cured in place pipe (CIPP), and more recently, high build polyurethane products. While these processes are readily available, to they have been used primarily for specialty applications and not on large gridded areas. However, as awareness to and experience with these materials increases, their use will be more consistent within the water pipe rehabilitation industry.

Pipe Bursting

Pipe bursting, which has been used in the water pipe rehabilitation industry since the 1980s, has gained significant technical advances within the last several years. The equipment has been continuously improved to increase production rates, pipe diameters, and length of possible bursting runs.

Until recently, bursting projects that involved service connections also required the installation of temporary water supplies during construction. However, Florida's Department of Environmental Protection (FDEP) has recently approved the use of prechlorinated pipe for bursting projects, thus eliminating the need for costly, time consuming temporary lines and the need for lengthy water quality testing following installation. Thus, customers are without water only for the time required to physically burst the main and restore service. Due to the relatively short service interruption time, no temporary service network is installed.

To facilitate the pipe bursting without establishing a temporary service network, a significant amount of planning and prepatory work must be completed. Prior to bursting existing pipe, the complete length of new main must be assembled on the ground surface. This new main is hyper-chlorinated, flushed, filled and sampled similar to any new main installation practices. Once the water in the new main is approved by the local regulatory agency, it is drained, sealed, and then pulled into place as the existing pipe is burst. This new main is then connected to the existing network, flushed and service connections are reinstated. Prechlorinated pipe bursting jobs in Orlando, Miami, Winter Haven, Jacksonville, and Ponte Vedra Beach, indicate that project costs may be reduced as much as 50 percent and that construction time lines can be shortened by as much as 90 percent, compared to traditional open-cut and replacement methods (Griffin, 2002).

Cured In Place Pipe

Cured in place pipe, (CIPP), a product frequently utilized for gravity sanitary sewer rehabilitation, is gaining recognition as a potable water pressure pipe renewal alternative. The CIPP process has had limited employed in Canada over the past several years. During these applications, equipment has been developed that can enter the pipe and reinstate service laterals, robotically, from within much like the process performed in sewer lines. However, the water service taps are generally much smaller than sewer laterals and therefore more difficult for the equipment to locate after lining. To reduce the time associated with reestablishing service connections, precise mapping of the service line locations prior to lining installation is essential. To further speed this process, future developments may include "locator" chips in the service connection plugs. A faster setting material is also under development.

Spray Applied Polyurethane

Polyurethane (PU or polyurea) is a spray applied lining material that is widely used in the United Kingdom but has very limited application in the United States to date. Within the United Kingdom, these materials have almost completely replaced the use of epoxy due to its speed of application and curing. The rapid set characteristics, combined with a different sampling protocol, enable contractors to routinely line 500 feet per day and maintain a lining rate of 10,000 feet per month. The use of this material in the United States would require a change to standard rehabilitation procedures and a rethinking of the disinfection and sampling regime practiced by most utilities in the United States and Canada and a change in regulatory requirements.

CHAPTER 2
DATA ACQUISITION

INTRODUCTION

The data necessary for this research project was provided by utilities actively involved in pipe lining activities. Participating utilities provided the necessary data by responding to survey questions, telephone interviews, and questionnaires. The results and conclusions from this body of data are thus biased toward the information provided by participating utilities.

UTILITY SURVEY

Survey Protocol

The utility survey questionnaire was completed during initial phases of the project . In general, the overall purpose of the survey was to identify utilities that were actively involved in pipe lining activities and willing to participate in the data collection efforts. To complete the survey and obtain the necessary information, the following steps were followed:

- Acquiring email distribution lists
- Generating survey questions
- Administering web base survey
- Tabulating survey results

Approximately 1,300 email addresses that represented an equal weighting of small, medium and large water utilities were obtained. However, it was determined that the utility contact information obtained did not include the appropriate person to answer technical questions regarding pipe lining operations. As such, a two staged contact method was implemented where the initial survey only requested the initial contact to identify the individual within their utility who is best positioned to answer technical questions. The initial contacts list provided approximately 150 focused references of individuals judged capable of answering technical questions. A second, more detailed survey regarding minimizing down time during pipe lining operations was distributed and 47 utilities completed the questionnaire.

The survey questions were designed to assess the scope and range of pipe lining experience and projects within each utility. Once the survey questions were finalized, the Urban Studies Institute at the University of Louisville administered the web-based survey. The initial contact survey was distributed on March 26, 2004 and the technical survey was distributed on April 26, 2004. Both surveys remained open for approximately 10 days and a reminder email was distributed half way through the response window. Upon conclusion of the survey, the results were tabulated and histograms created to summarize data.

Survey Results

The survey consisted of 12 layered questions designed to extract technical data about a utilities pipe lining experience. Results of the survey are displayed in Figures 2.1 to 2.12:

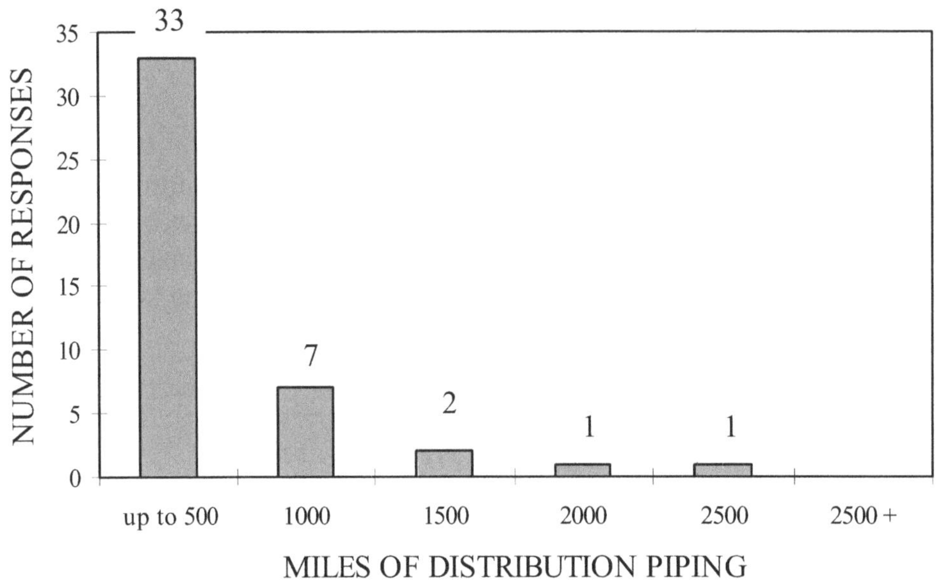

Figure 2.1: Approximate length of transmission and distribution piping maintained by utilities responding to survey questions

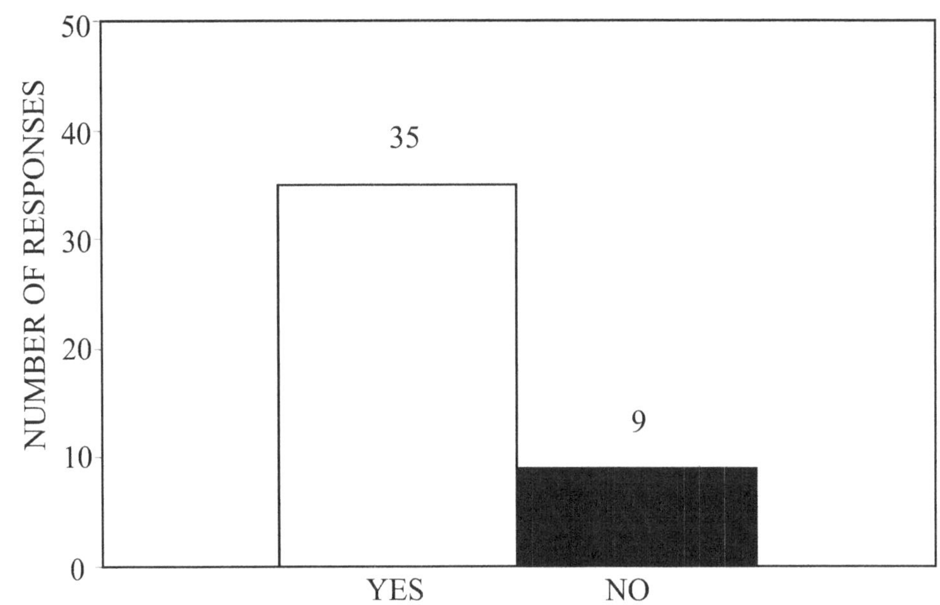

Figure 2.2: Number of utilities with main replacement / rehabilitation program responding to survey questions

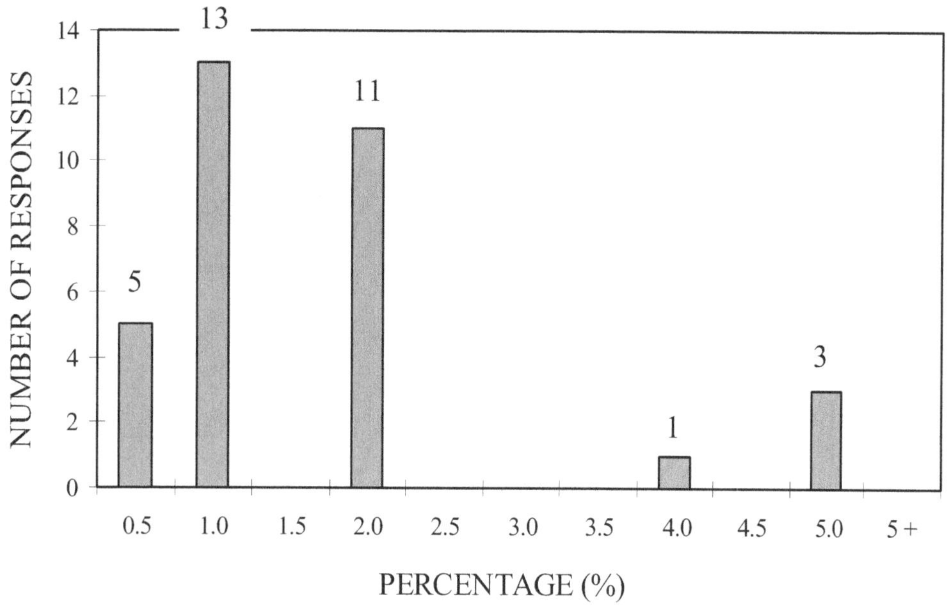

Figure 2.3: Percentage of distribution system *replaced* annually by utilities responding to survey questions.

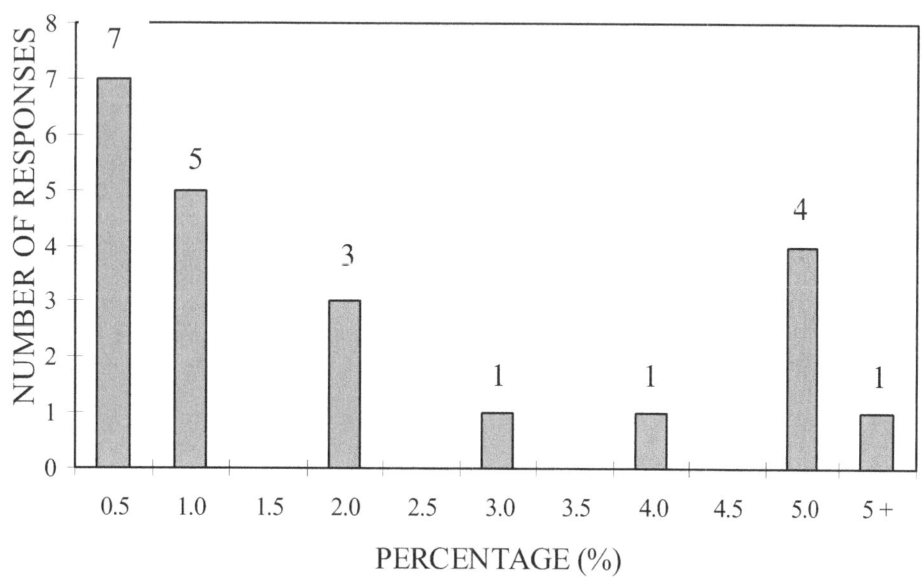

Figure 2.4: Percentage of distribution system *rehabilitated* annually by utilities responding to survey questions.

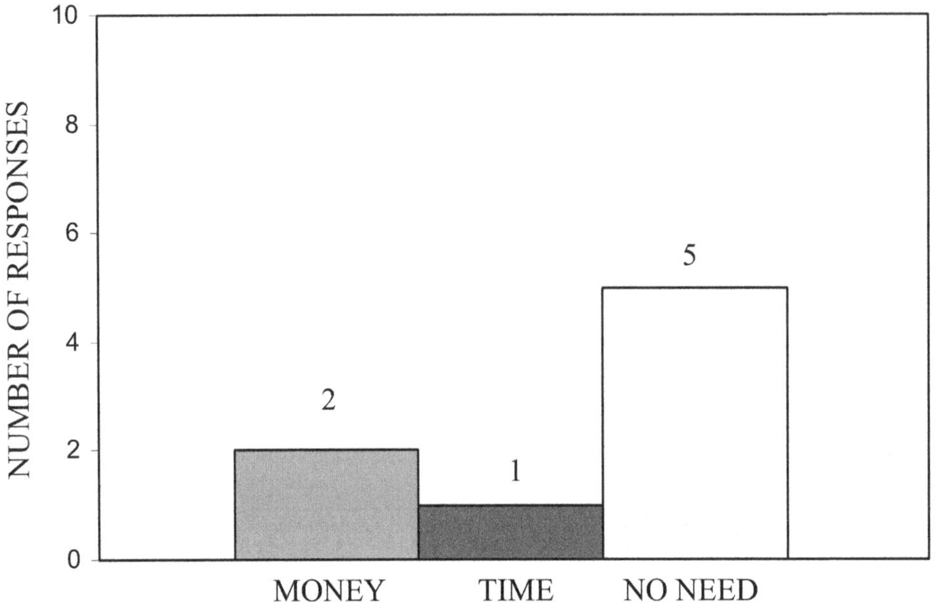

Figure 2.5: Reasons for not initiating a rehabilitation program provided by utilities responding to survey questions.

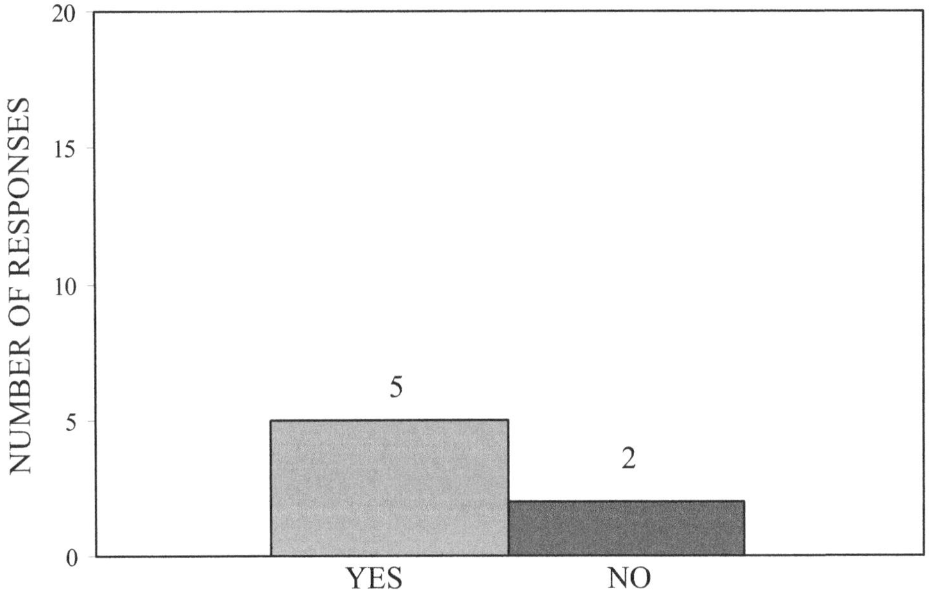

Figure 2.6: Number of utilities in the survey that plan to initiate a replacement-rehabilitation program.

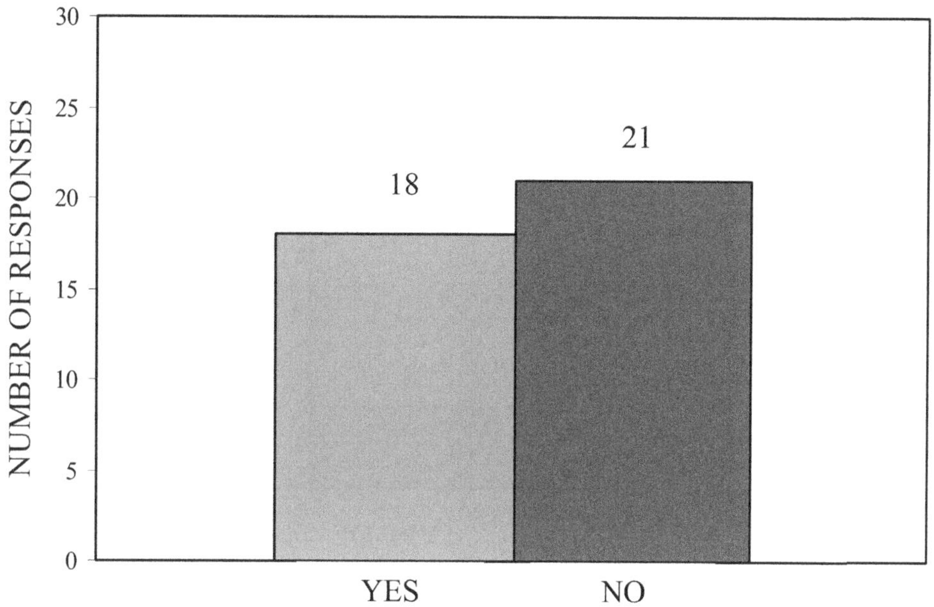

Figure 2.7: Number of utilities responding to the survey that regularly supply temporary water to customers during emergency activities.

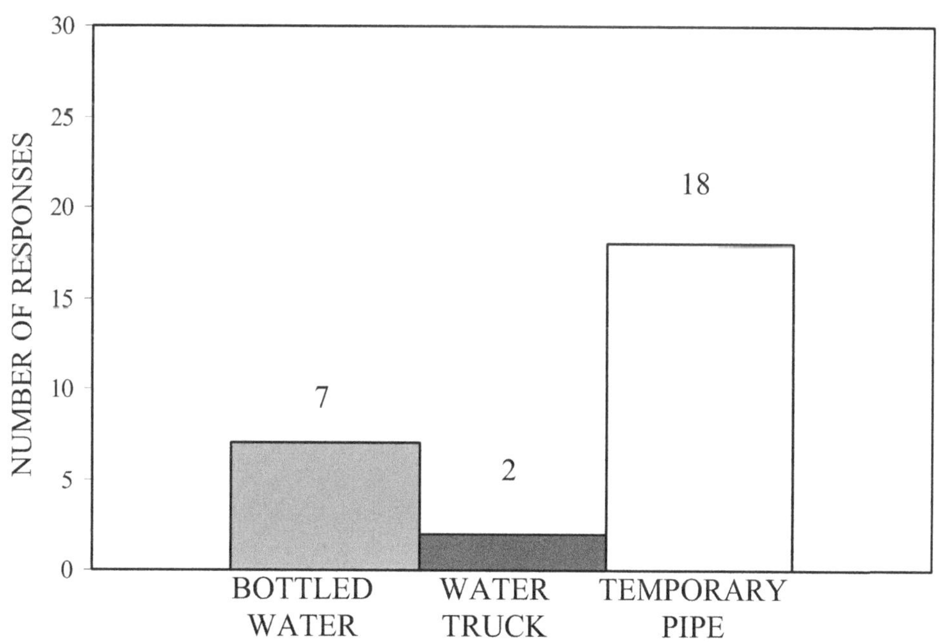

Figure 2.8: Methods typically used by utilities responding to the survey to supply temporary water during emergency outages

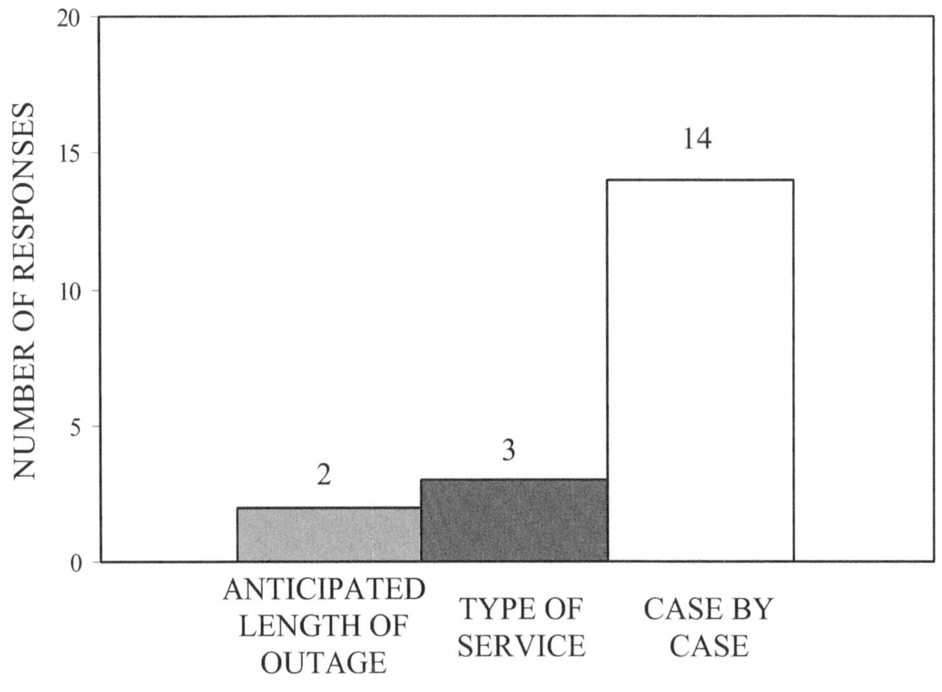

Figure 2.9: Procedures utilities responding to the survey use to determine appropriate temporary supply methods for use during emergency outages

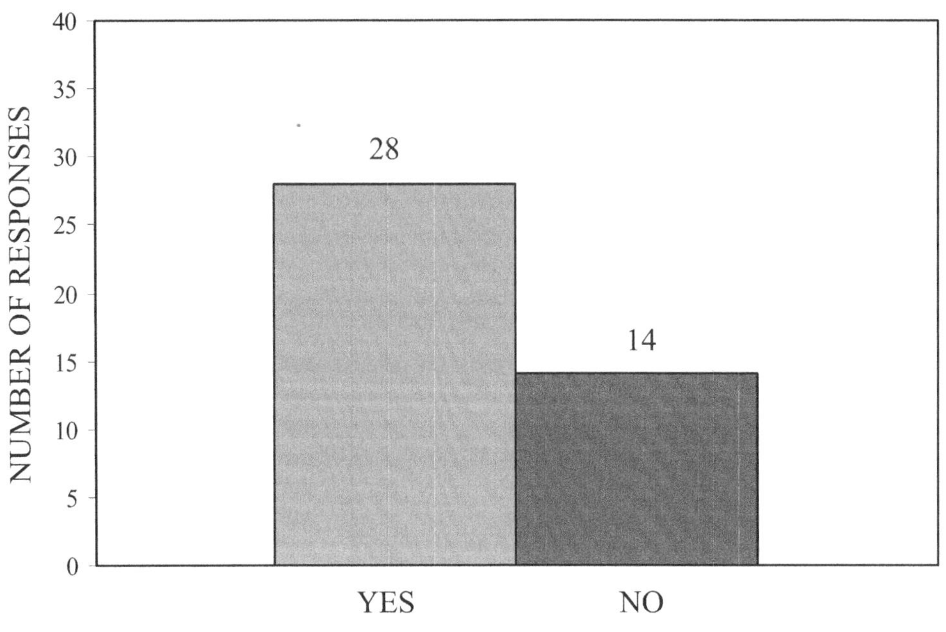

Figure 2.10: Number of utilities responding to survey that routinely provide temporary water to customers during planned maintenance and/or capital improvement activities.

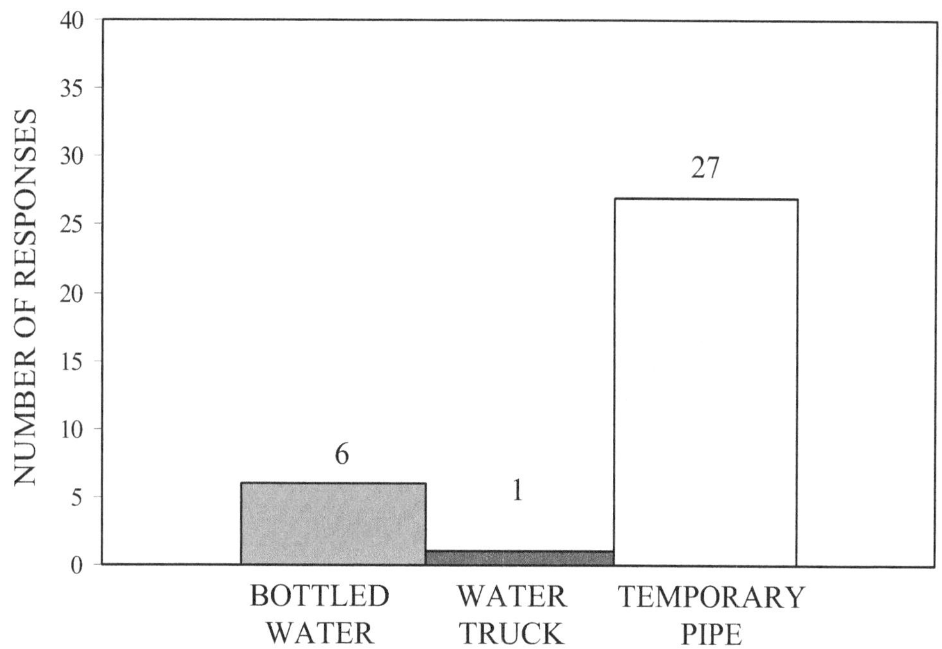

Figure 2.11: Methods used by utilities responding to the survey to supply temporary water during planned maintenance outages.

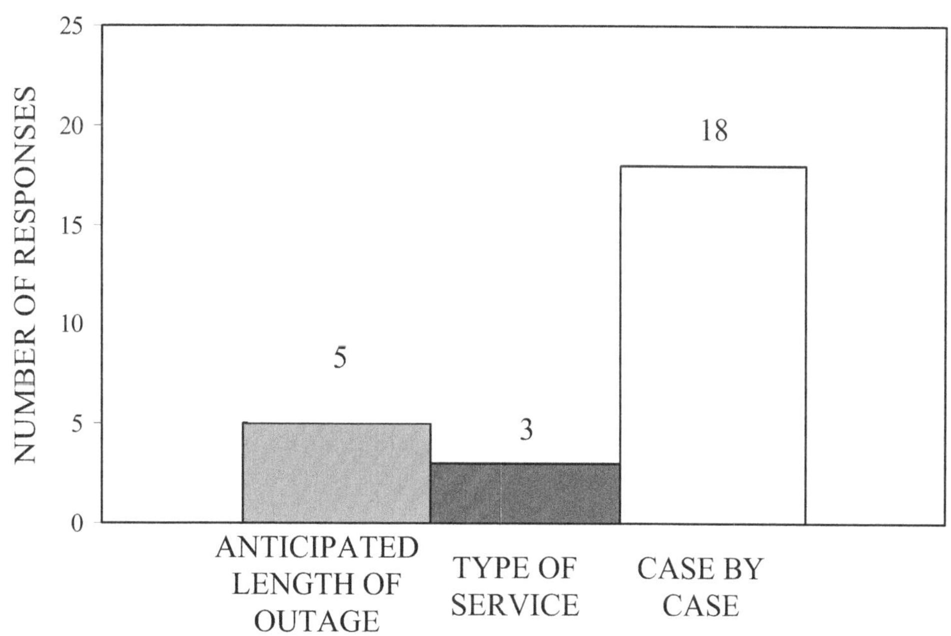

Figure 2.12: Method that utilities responding to the survey use to determine appropriate temporary water supply methods during planned maintenance outages

MINIMIZING DOWNTIME FIELD DATA

Data Acquisition

Once the survey was complete, the utilities that indicated they had an active lining program were contacted. Each utility was asked if they would collect data during their lining program such that best management practices could be identified. Of the 35 utilities with an active main line rehabilitation and replacement program, the following utilities agreed to participate in the data collection:

- Anne Arundel County, MD; Department of Public Works
- Baltimore, MD; Baltimore Metropolitan Water Distribution System
- Charlotte, NC; Charlotte-Mecklenburg Utilities
- Nashville, TN; Metro Water Services
- Louisville, KY; Louisville Water Company
- Edgewood KY; Northern Kentucky Water District
- Omaha, NE; Metropolitan Utilities District
- Cincinnati, OH; Greater Cincinnati Water Works
- Cleveland, OH; City of Cleveland, Division of Water

Each participating utility received a data collection form. The data acquisition form was organized to obtain critical time information for both the pipe lining and temporary water supply activities. Both man-hours and dates of occurrence information were collected in an effort to identify possible efficiencies. The data sheet provided to the utilities is displayed in Figure 2.13.

The participating utilities provided information specific to their pipe lining activities. The utilities were requested to record data in individual city block segments to enable a consistent basis for comparison. A summary table containing a listing of the lining method and number of blocks included in the data collection process, and the total length of lining is displayed in Table 2.1. Of the utilities contacted and agreeing to participate, none were involved in slip lining CIPP activities. Thus, critical data regarding these lining techniques could not be assessed.

Data Acquisition Results

The following sections provide an overview of the materials, methods and manpower for both temporary pipe and lining activities as reported by the participating utilities. While the data and trends observed are believed to be reflective of the industry as a whole, the analyses and conclusions will obviously be biased towards participating utility operations.

Temporary Pipe Installation

Providing temporary water service during pipe lining activities is a significant component on the lining process. The data reflects a variety of materials, methods and regulations that pertain to temporary pipe installation.

WATER MAIN LINING PROJECT INFORMATION
(AwwaRF Project 2956)

LOCATION: ______________________________
(Street Name)

FROM: ______________________________
(Intersecting Street)

TO: ______________________________
(Intersecting Street)

EXISTING PIPE TYPE: ______________________________
(cast iron, steel, ductile iron, PVC, AC, PCCP, other)

LINING METHOD: ______________________________
(cement mortar, epoxy, urethane, HDPE, CIPP, other)

EXISTING MAIN INFORMATION:

Diameter [ ] Length [ ]

No. of Service Connections [ ] No. of Fire Hydrant Connections [ ]

Date:		Man-hours:	
Dewatered	[ ]	Dewatering	[ ]
Cleaned	[ ]	Cleaning	[ ]
Lined	[ ]	Lining	[ ]
Chlorinated	[ ]	Chlorinating	[ ]
Flushed	[ ]	Flushing	[ ]
Sampled	[ ]	Sampling	[ ]
Approved	[ ]	Approving	[ ]
Reinstated	[ ]	Reinstating	[ ]

BYPASS PIPE INFORMATION:

BYPASS PIPE TYPE: ______________________________
(PVC, steel, HDPE, rubber hose, other)

Diameter [ ] Length [ ]

Date:		Man-hours:	
Installed	[ ]	Installing	[ ]
Chlorinated	[ ]	Chlorinating	[ ]
Flushed	[ ]	Flushing	[ ]
Sampled	[ ]	Sampling	[ ]
Approved	[ ]	Approving	[ ]
Services Connected	[ ]	Connecting Services	[ ]
Removed	[ ]	Removing	[ ]

Total Number of Driveways Crossed [ ]

Number of Driveways Ramped w/Asphalt [ ]

Number of Driveways Ramped w/Rubber Ramps [ ]

Footage Buried in Street Crossings [ ]

Number of Customer Complaints [ ]

Number of Temporary Pipe Breaks [ ]

Please forward completed forms to:

Center for Infrastructure Research
Attn: Thomas D. Rockaway, Ph.D., P.E.
101 W. S. Speed Building
University of Louisville
Louisville, Kentucky 40292

Figure: 2:13: Main lining data acquisition sheet provided to utilities participating in the study

Table 2.1:
Summary of data collected from participating utilities during pipe lining operations

Utility Location and Name	Abbr.	Number of Sites	Lining Length (ft)	Lining Method
Anne Arundel County, MD Department of Public Works	AACDW	10	8,699	Cement Mortar
Baltimore, MD Baltimore Metropolitan Water Distribution System	Baltimore	3	5,245	Cement Mortar
Charlotte, NC Charlotte-Mecklenburg Utilities	Charlotte	5	5,625	Epoxy
Nashville, TN Metro Water Services	Nashville	11	11,343	Cement Mortar
Louisville, KY Louisville Water Company	LWC	8	2,985	Cement Mortar
Edgewood KY Northern Kentucky Water District	NKWD	7	10,856	Epoxy
Omaha, NE Metropolitan Utilities District	Omaha	6	4,775	Cement Mortar
Cincinnati, OH Greater Cincinnati Water Works	GCWW	14	6,209	Cement Mortar
Cleveland, OH City of Cleveland Division of Water	Cleveland	10	16,903	Cement Mortar

Temporary Pipe Material, Length and Percent Used: During the pipe lining process, the participating utilities employed a variety of materials to convey temporary water to their customers. Based on the data collected PVC, HDPE, and nitrile rubber were the most common temporary piping materials, with one utility employing a very minor amount of steel pipe. The steel pipe is included in the charts and figures, but due to the limited amount significant discussions or conclusions regarding steel pipe could not be provided.

A total of 146,418 feet of temporary pipe was installed as part of this study. PVC was the most commonly used material accounting for approximately 59%, or 86,500-feet. HDPE pipe and nitrile rubber accounted for 25% (37,100-feet) and 15% (21,900-feet) of the temporary pipe installed, respectively. Approximately 900 linear feet of steel pipe, accounting for only 0.6% of the total temporary pipeline was installed. Figure 2.14 shows the amount of temporary pipeline installed displayed by pipe material type and as a percent of the total pipe installed.

Length of Temporary Pipe Required: A water main in operation will typically provide service to both sides of a street. When an existing water main is removed from service it is difficult to provide temporary above ground service using the same below ground distribution

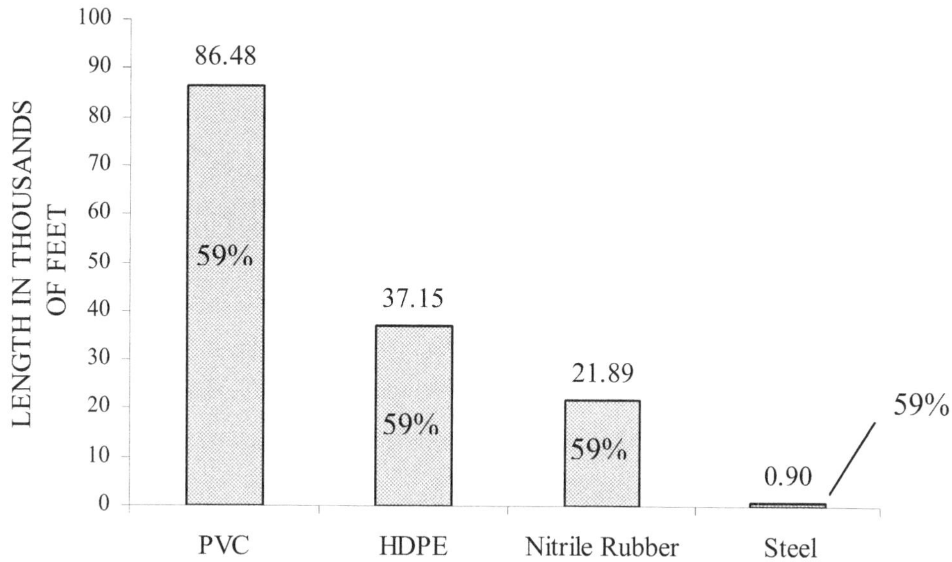

Figure 2.14: Temporary pipe installed totaled by material type

pattern without significantly disrupting traffic and pedestrian flow. As such, the amount of temporary pipe that must be installed is significantly greater than the footage of main removed from service.

The following are the actual footages of temporary pipe used on the most recent projects completed by project participants. The temporary pipe lengths are shown as a function of the total project scope without regard to pipe diameters. The amount of pipe lined within each utility ranged from 16,903 linear feet in Cleveland, Ohio to 2,985 linear feet in Louisville, Kentucky. The ratio of temporary pipe to length of line pipe ranged from a high of 2.4 feet in Nashville, Tennessee, on an 11,343 foot project to a low of 1.3 feet of temporary pipe in Cincinnati, Ohio on 6,209 linear feet project. The average for the 9-utilities involved in this study was 2 linear feet of temporary pipe for every foot of lined pipe in the project scope. Figure 2.15 shows the total length of temporary pipe installed and the amount of existing pipe renewed by each utility.

Common Types of Temporary Pipe: The diameter of temporary pipe used to provide service to customers depends on site factors including required velocity, flow quantity, fire protection and length of temporary. However, larger diameter temporary pipe is generally more expensive, difficult and time consuming to install, thus there is an incentive for utilities to minimize the size of the bypass piping.

The utilities surveyed accommodated customer demands by installing temporary bypass pipe in sizes ranging from 2-inches to 6-inches in diameter. The majority of the bypass pipe installed was 2-inch diameter, accounting for approximately 49% of the total footage (76,200 feet). Additionally, approximately 36% of 4-inch and 15% of 3-inch diameter temporary pipe was installed (52,400-feet and 2,900 feet, respectively). The data also included a very small amount (approximately 1,100 feet) of 6-inch diameter pipe which accounted for less than 1% of the temporary pipe installed. Figure 2.16 shows temporary pipe by diameter, as a percent of the total pipe installed. Figure 2.17 shows the same data by total length installed by pipe diameter.

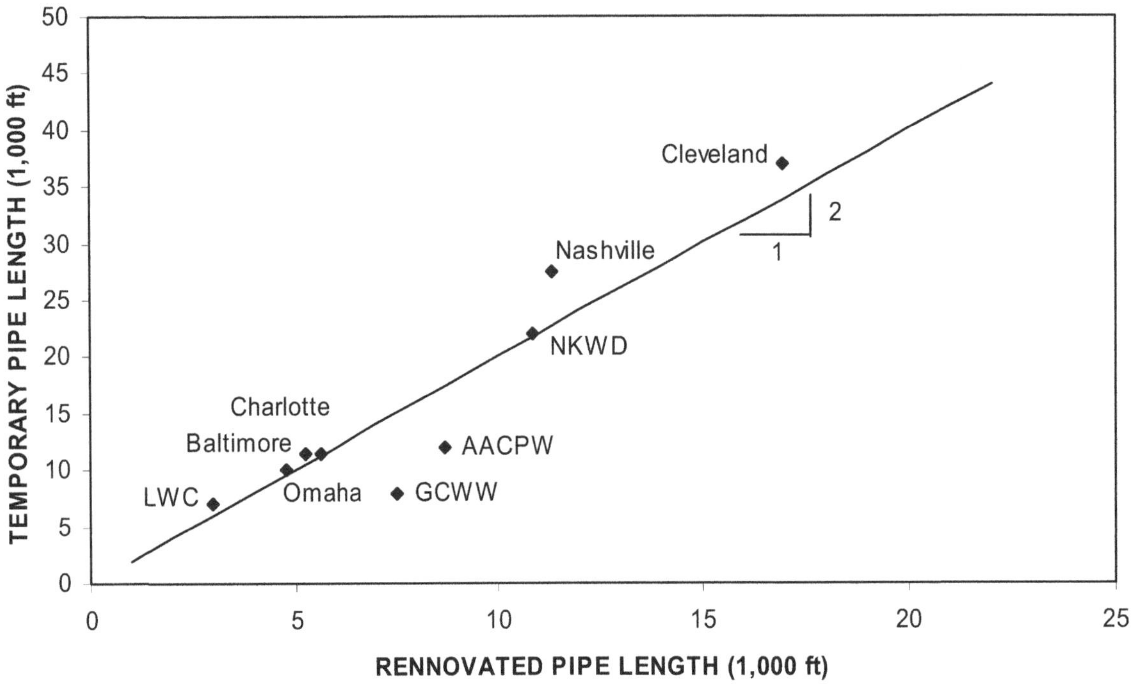

Figure 2.15: Comparison of length of temporary pipe installed to the length of existing main renewed

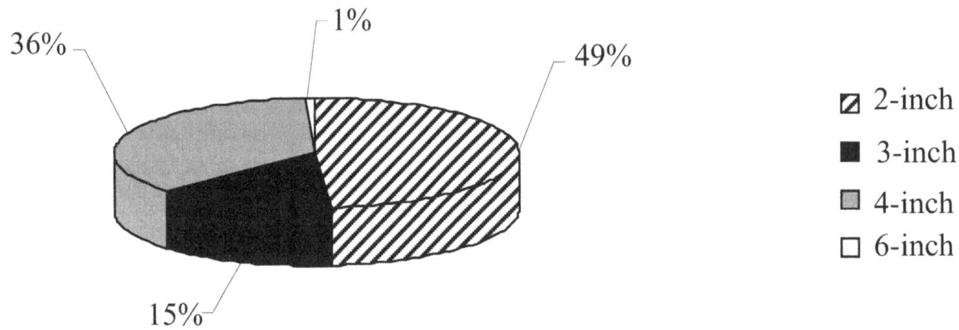

Figure 2.16: Temporary pipe by diameter shown as a percent of the total installed

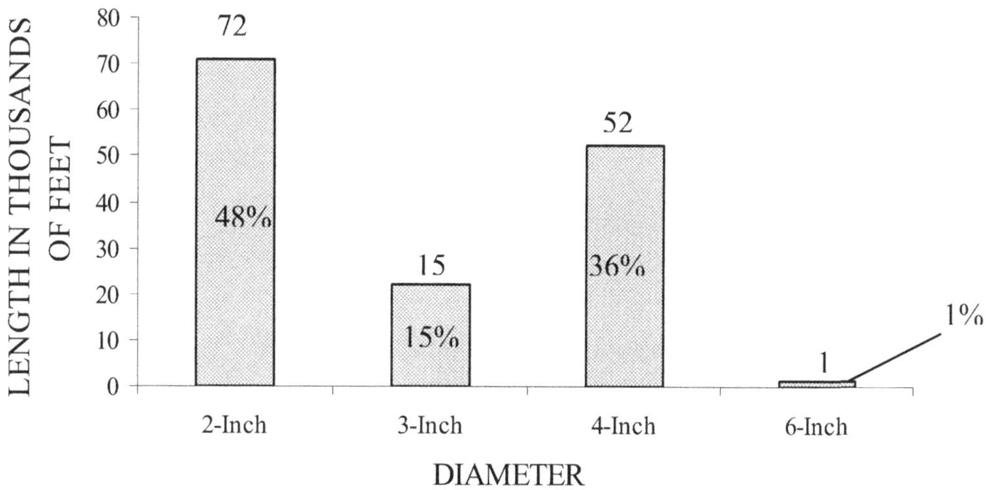

Figure 2.17: Length of temporary pipe installed, by diameter

Temporary Pipe Manpower Requirements: Manpower utilization records can provide a consistent means to compare levels of effort between separate tasks and to identify efficient operations within a task. As such, manpower utilization was collected for individual tasks for both the activities required to install and maintain a temporary pipeline network and the tasks involved in the renewal of the existing pipeline.

Within a pipe rehabilitation program, the two primary components of the work are the temporary water operations and the actual pipe lining process. To assess the relative level of effort devoted to each of these categories, the total number of man hours devoted to rehabilitation was compared to the total number of man hours related to installing the temporary pipeline. As shown in Figure 2.18, approximately 41% of the recorded man hours were devoted to temporary pipe operations while 59% was directed towards actual pipe rehabilitation. These figures, however, account only for the hours devoted to lining and temporary pipe. They do not include the hours necessary to excavate access pits, backfill, and restoration activities. Nevertheless, temporary pipe installation and maintenance is a significant percentage of the lining process.

Manpower Requirements for Temporary Pipe Tasks: Within the general temporary pipe and rehabilitation activities, the manpower records were further subdivided according to individual tasks. The data collection sheets directed each utility to document the total man-hours spent on temporary pipe activities including: installation, chlorination, flushing, sampling, approving, connecting services, and removing. In addition, for rehabilitation of the existing pipe man hour requirements were recorded for specific tasks including: dewatering, cleaning, lining, chlorinating, flushing, sampling, approving, and reinstating. Completion of each of these tasks was also tracked by calendar days so as to differentiate between tasks that may be manpower driven and those that may be regulatory driven. As an example, chlorination will take relatively few man hours to apply disinfectant and retrieve a sample, however, the 24-hour contact time (CT) controls the task time. Thus, additional workers applied to the task will not accelerate the process.

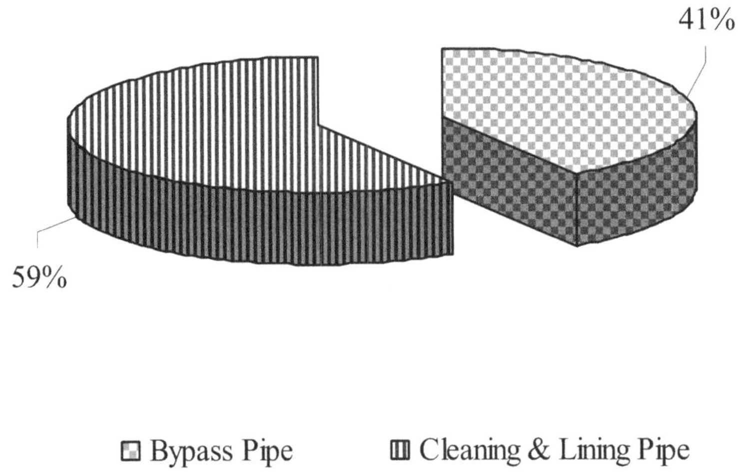

Figure 2.18: Distribution of manpower for lining tasks and temporary pipe tasks as a percent of the total man-hours

Of the activities surrounding temporary pipeline, the physical installation of piping materials required the greatest number of man hours, average 36% of the manpower resources. Connecting services, removing, and chlorinating consumed 23%, 19%, and 12% of the manpower resources, respectively. The remaining three tasks, flushing, sampling, and approval, combined accounted for 10% of the required manpower. Figure 2.19 illustrates the manpower distribution, by task for installation and maintenance of a temporary water system.

Of the eight tasks monitored as a part of the cleaning and lining activities, reinstating the pipe to service after lining required the greatest amount (46%) of the total manpower. The cleaning and lining combined for a total of 35% of the necessary manpower and the remaining five tasks combined (chlorinating, flushing, sampling, approving and removing) accounted for 19% of the necessary manpower. Figure 2.20 shows the manpower distribution, by task, for cleaning and lining water mains.

Service Density and Fire Hydrant Considerations: The number of services, fire hydrants, street crossing and driveways within a rehabilitation project section has a significant effect on the project time. As the number of obstacles or system interruptions increases, additional manpower will be required to establish and maintain connections.

For the project data recorded, service density ranged from a high of one service for every 10 linear feet of existing pipe on a 300-foot segment, to a low density of one service attachment on a 244-foot segment. The average domestic service density among all 74-pipe segments studied was one service for every 37 linear feet of pipeline. One 278-foot segment of 8-inch diameter relined pipe had no service connections.

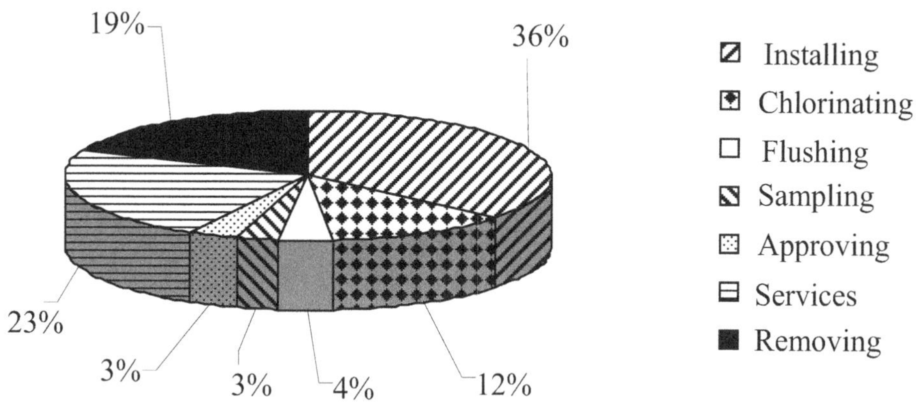

Figure 2.19: Manpower distribution by task to install and maintain temporary pipeline.

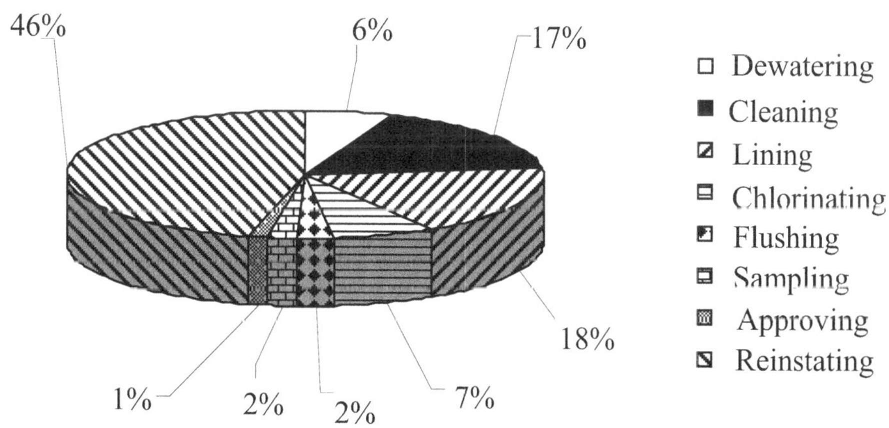

Figure 2.20: Manpower distribution by task to clean and line existing pipe

Fire hydrant densities ranged from a high of one hydrant per 100-linear feet of main to a low of one hydrant per 1,500 linear feet of main. The average distance between fire hydrants on the 75-pipe segments studied was 501 linear feet. However, 13-segments ranging in length from 155 linear feet to 600 linear feet had no hydrant attachments. The number of hydrants per rehabilitation section becomes important when temporary fire hydrants must be installed for every hydrant removed from service during pipe rehabilitation.

Buried Temporary Pipe: Street crossings and large commercial driveways typically require excavating enough roadway material to bury the temporary pipe flush with the road surface. While expensive and time consuming, this activity is believed to have a positive effect in reducing both the number of pipe breaks and customer complaints. As such, data was collected to determine the typical amount of temporary pipe line that is buried in paved areas. Based on the data collected, the percent of buried pipe to total pipe ranged from a low of 1% in Charlotte and Northern Kentucky to a high of 11% in Louisville. Figure 2.21 shows the total length of temporary pipe installed by each utility as well as the lengths at grade and buried, and the percent of the total length that was buried.

Driveway Crossing Techniques: Customer interaction with the temporary pipe at their driveway is believed to be one of the controlling factors regarding breaks and complaints. A number of methods are available for use at driveway crossings to allow residential vehicular traffic to traverse the pipe without damage and with minimum annoyance. Some of the more common driveway crossing methods include: prefabricated rubber ramps, temporary asphalt (or cold patch ramps), wooden ramps and trenching and burying the pipe. In some instances, as is the case with nitrile rubber pipe, no ramps or specialized crossing devices are necessary. Of the ramping methods used, rubber ramps were the most common as they were used at 68% of the study driveways crossings. Twenty-two percent of the driveways were ramped using temporary asphalt and 10% had no ramps. Figure 2.22 shows the distribution of driveway ramping techniques. Figure 2.23 show the distribution of techniques used by study participants.

Pipe Lining Operations

The physical characteristics of the lined pipe to be lined and the selected lining method will directly affect the amount of time, expense and effort required to complete the rehabilitation effort. While the utilities may not be able to control the site conditions or physical attributes of the pipe, these characteristics can be correlated to project duration and manpower. Additionally, the rehabilitation method selected should affect the overall installation rates. The collected data shows the variations for pipe size, type, and rehabilitation method for the project sites.

Existing Pipe Diameters: Within older cities, 6-, 8-, and 12-inch diameter mains make up the vast majority of the distribution system and also make up the majority of the mains requiring renewal. Figure 2.24 shows that of the 72,640-feet of cleaned and lined pipe included in the study, 6-inch, 8-inch, and 12-inch diameter pipes accounted for 92% of the total footage. Of the remaining, four-inch diameter pipe accounted for 6% and 10- and 20-inch accounted for 1% each. Figure 2.25 illustrates the length of various pipe diameters lined by each utility.

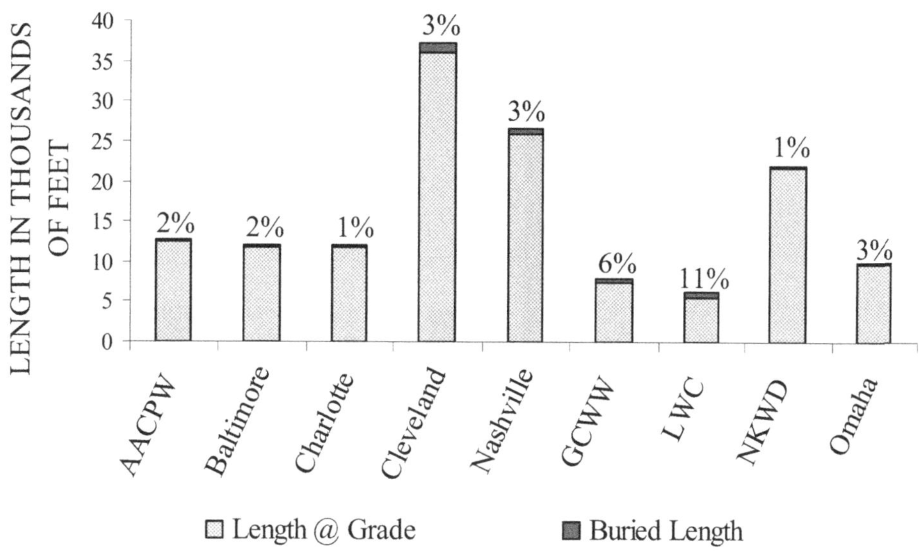

Figure 2.21: Typical temporary pipe footage and burial length

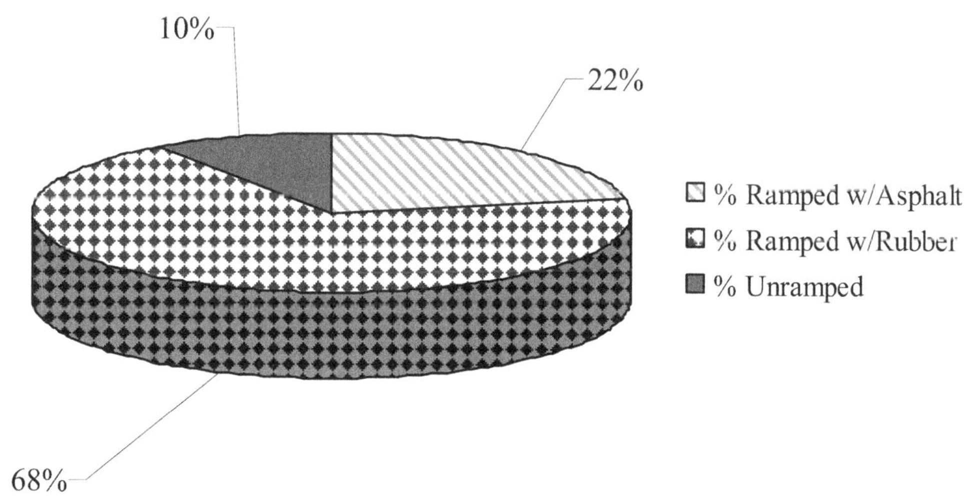

Figure 2.22: Distribution of various driveway ramping techniques

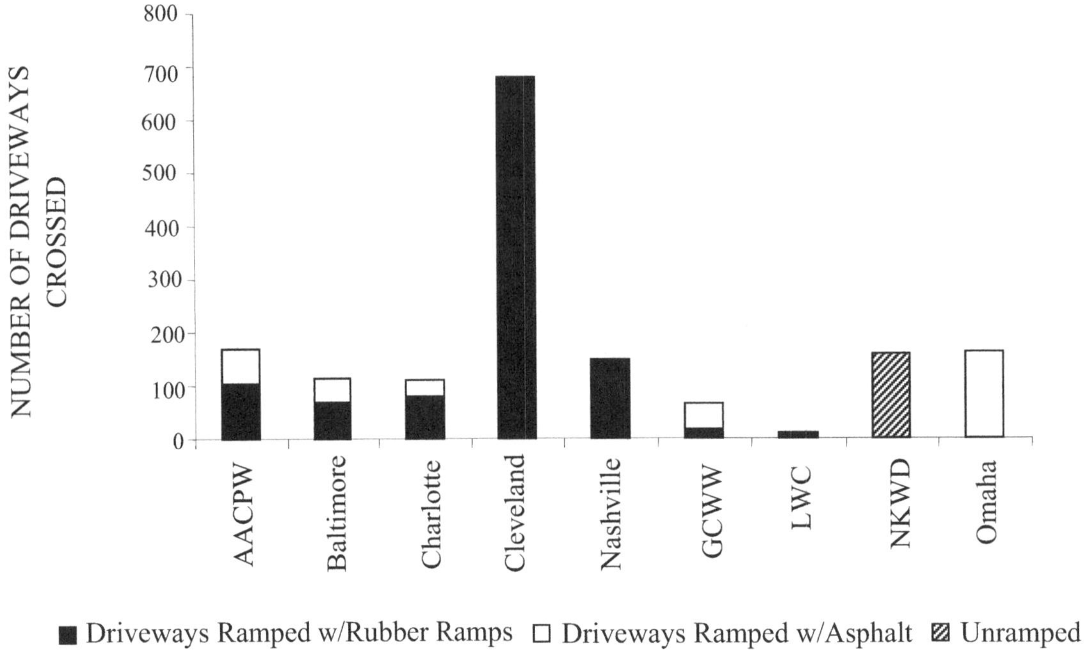

Figure 2.23: Distribution of driveway ramping technique, by utility

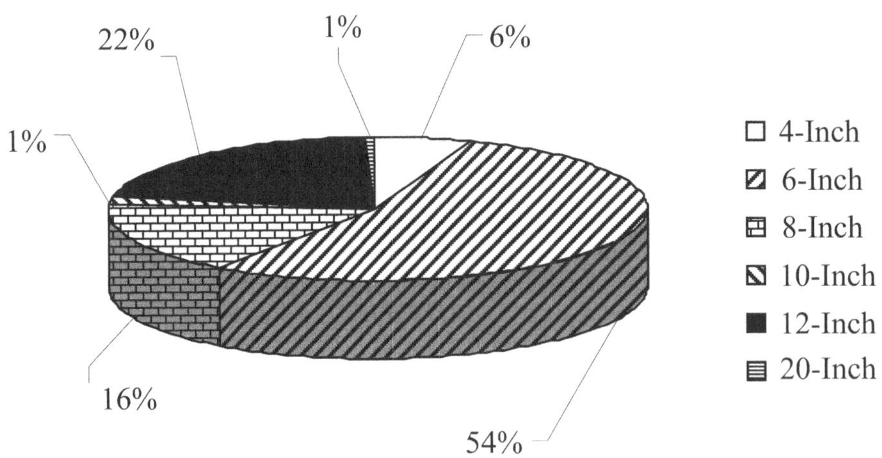

Figure 2.24: Variety of pipe sizes cleaned and lined (as reported by utility participants)

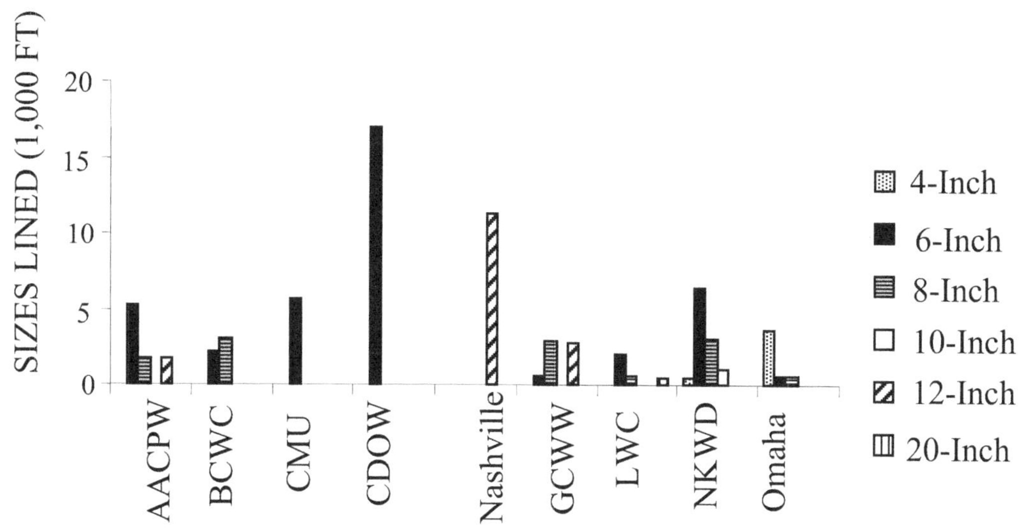

Figure 2.25: Length and diameter of pipes cleaned and lined (as reported by participating utilities)

Manpower Required to Rehabilitate Existing Mains: In an effort to determine manpower requirements for the pipeline rehabilitation tasks the existing pipes are categorized by diameter. In general, for most pipeline rehabilitation tasks, larger diameter pipelines require a larger number of man-hours (on a per foot or per unit basis). An exception to this statement would be the sampling and approval tasks which are independent of pipe size. Within Figure 2.26 the total man-hours for all rehabilitation tasks were tabulated by pipe diameter. The 10- and 20-inch diameter pipelines were not included within this tabulation due to the limited number of segments within our study.

The data set contained 5 segments of 4-inch pipe totaling 4,145 feet; 37 segments of 6-inch main totaling 39,415-feet, 11 segments of 8-inch pipe, totaling 11,840-feet, and 19 segments of 12-inch pipe totalling15,797-feet. The total man-hours for each pipe diameter was calculated and normalized to 500-feet, approximately one-city block. The average number of man-hours per the normalized 500-feet segment was; 37-hours; 51-hours, 98-hours, and 244-man-hours for the 4-, 6-, 8- and 12-inch diameter pipes, respectively.

Common Lining Materials: The two commonly used lining materials in the United States are cement mortar and epoxy coating. While other water lining methods are available, none was used by study participants. The 74- cast iron pipe segments included in this project totaled 72,640 linear feet. Of that amount 77% (56,159 linear feet) was cleaned and lined with cement mortar. The remaining 23% (16,481 linear feet) was cleaned and coated with a spray applied epoxy material. Of the nine utilities surveyed, Northern Kentucky Water District and Charlotte Mecklenburg Utilities were the only utilities with specifying an epoxy coating program. Figure 2.27 shows the distribution, by percent of cement mortar usage and epoxy resin.

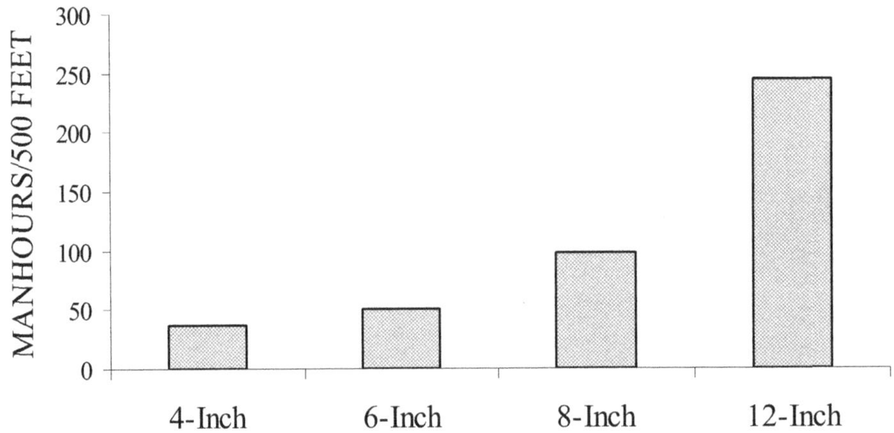

Figure 2.26: Average number of man-hours to cleaned and lined 1,000-feet of pipe, by size

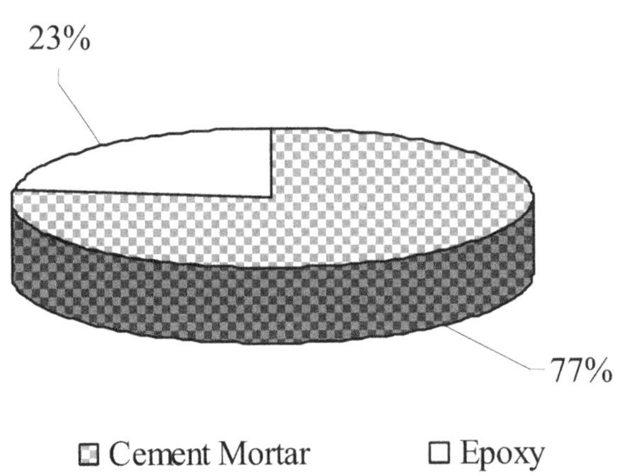

Figure 2.27: Water main lining materials within the sample study.

Manpower Requirements Based on Lining Material: With the available data, a manpower comparison could be made between the pipe segments cleaned and cement mortar lined and those cleaned and epoxy coated. Only 6- and 8-inch diameter pipes were included in the comparison as those are the only pipe sizes within the data sets that include both lining methods. The data contained approximately 36,250 linear feet of 6-inch and 8-inch pipe cleaned and cement mortar lined. This effort required approximately 2,150 man-hours, or about 30-man hours per 500-linear feet. Cleaning and epoxy coating required approximately 834-man-hours to clean and coat about 15,000 linear feet of pipeline, or 28-man hours per 500-segment as shown in Figure 2.28.

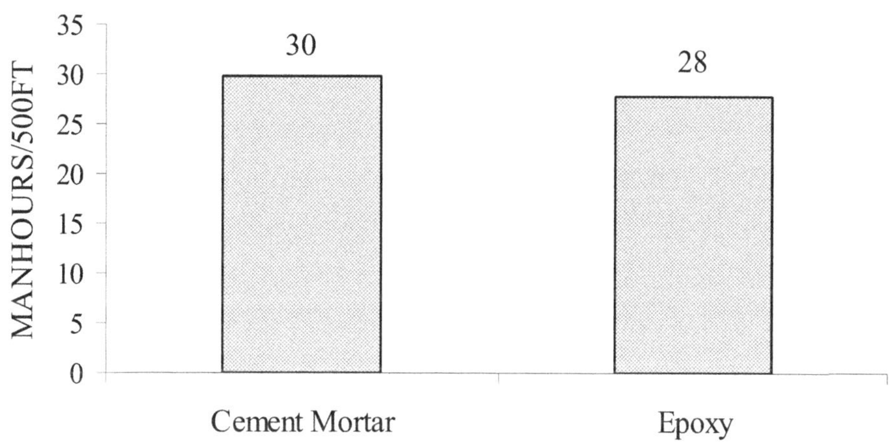

Figure 2.28: Average number of man-hours to clean and line 500 feet of 6-inch and 8-inch pipe, by lining material

CHAPTER 3
DATA ANALYSIS

INTRODUCTION

The data collected from the participating utilities provided the mechanism to critically review current pipe lining practices. The data were analyzed to define current state of practice and then further analyzed to identify efficient operations. Efficient operations were identified through a statistical analysis of man hour requirements and project durations for each task or project segment. In this fashion, the best management practice recommendations are based on actual data from utilities actively engaged in the lining process. However, the trends and recommendations are biased towards the activities and procedures performed by the participating utilities.

The data collected focused on gathering critical information regarding the pipe lining process. For the analysis, the information was subdivided into three primary analysis categories including:

- Temporary pipe installation
- Cleaning and lining task analysis
- Customer complaints

Listed below is a discussion of the observed trends and best management practices for each of these areas as determined by the data collected and statistical procedures implemented in the analysis. Ultimately, the combination of these factors play a critical role in determining the overall time and cost requirements necessary to install temporary pipe within each city.

The analysis has normalized the collected data to a typical base unit to enable a comparison of productivity rates between utilities. The base unit for comparing existing waterline is 500 feet. This represents the estimated length of a city block in older developments, which are likely to contain unlined cast iron mains. The base unit for comparing temporary pipe is 1,000 feet. This unit was selected based on Figure 2.15 of Chapter 2 that determined for every foot of existing main renewed, there are 2 feet of temporary pipe installed.

TEMPORARY PIPE INSTALLATION

Reviewing the various materials available for use as temporary water supply was a primary objective of this project. The data collected from participating utilities was used to evaluate the variety of pipe types for both durability and ease of installation. This was accomplished by tracking the; man hours and materials required to installing, maintaining, and removing the temporary pipelines; the number of breaks on each segment of temporary pipeline; and the number of customer complaints during the time the temporary pipeline was in place. The following discussion is intended to offer insight into the potential of each product as well as manpower requirements to help in developing budget-level construction estimates, and offer actual historical construction performance data.

Temporary Pipe Manpower Requirements

Pipe Material Selection

It is important to understand the various elements involved with establishing temporary water supply lines as they directly affect budgets and schedules of pipe rehabilitation projects. A variety of temporary pipe materials and sizes are available, and each has corresponding benefits and limitations.

Within our study, the three temporary pipe materials used by participating utilities were PVC, HDPE, and nitrile rubber. Installation rates for these materials ranged from 4 to 45 man hours per thousand feet. As discussed in Chapter 2, PVC was the most common temporary pipe material. However, when assessing the length of time necessary to install a unit length of pipe, nitrile rubber appeared to be significantly more efficient (Figure 3.1).

The relatively rapid installation rate for the 3-inch nitrile material is due to it's flexibility, durability and ease of use. This material is transported on large trailers containing several hundred feet of hose on spools, or hose reels (Figure 3.2). Installation of the hose consists of securing the hose end at the beginning of the project and pulling the trailer to the other end laying the hose out as the trailer moves down the street. Service connection tees are then installed at the coupling locations as needed to provide water supply to individual services (Figure 3.3). The hose is extremely flexible and requires few if any fittings to negotiate turns and radii (Figure 3.4). The hose is installed across residential drives without the use of ramps unlike PVC and HDPE material (Figure 3.5). Of the participating utilities that provided data, only NKWD employed the use of nitrile temporary water pipe.

The diameter of temporary pipe selected for a project site will also have a significant affect on the required installation time. Often, the diameter of temporary water pipe will be controlled by local jurisdictions and will depend on factors including: availability of water for fire fighting in the area surrounding the project, local regulations, local flow demands, and pipeline network configuration. For example a well gridded pipeline network, available fire hydrants, and an excellent relationship between the utility and the local fire department allowed

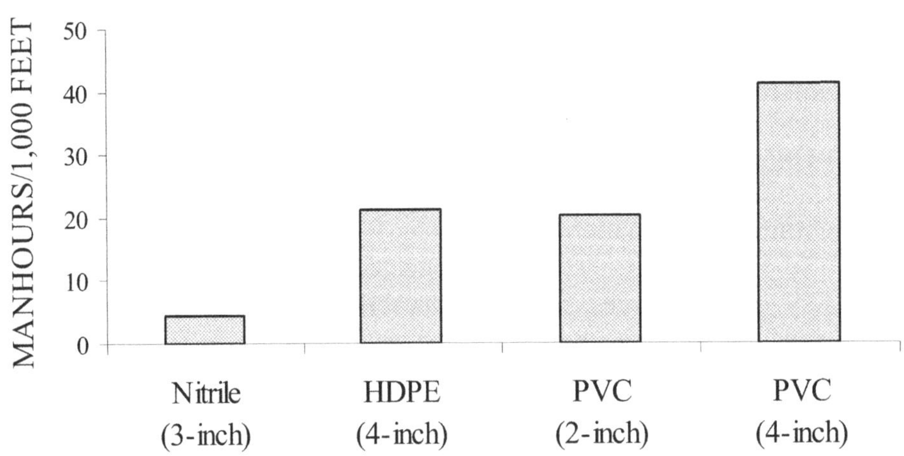

Figure 3.1: Average man hours feet/1,000 feet of temporary pipe installed

Figure 3.2: Nitrile rubber on hose reel

Figure 3.3: Service tee installed in nitrile rubber temporary supply system

Figure 3.4: Nitrile rubber temporary negotiating radius without fittings

Figure 3.5: Nitrile rubber crossing driveway without ramps

LWC to install primarily 2-inch diameter temporary lines and only a minimal amount of 4-inch. Additionally, only in the most extreme instances was it necessary to install temporary fire hydrants. In contrast, Cleveland requires a temporary hydrant for each existing hydrant taken out of service and a minimum 4-inch diameter pipe to service these hydrants.

To assess the affect of pipe diameter on installation rates, the PVC pipe data was further subdivided into 2-inch and 4-inch components as it was the only pipe type that had data for both diameters. The man hour comparison is based on 37,563 linear feet of 2-inch PVC installed in 25 segments and 42,918 linear feet of 4-inch PVC installed in 28 segments. Seven of the nine utilities that participated in the study used PVC pipe for temporary water supply and four of these used both 2-inch and 4-inch diameters. The installation rate data shows that the 2-inch pipe was installed at a rate of 20.3-man hours per 1,000-feet and the 4-inch was installed at a rate of 42.7-man hours per 1,000-feet. Street crossing, buried footage, driveway crossings, etc. were not taken into account when calculating these averages.

Service Density and Meter Location

The total number of services to connect to a given segment of temporary pipeline directly affects the amount of manpower devoted to that task. As the number of services per segment increases, so does the amount of time necessary to complete the installation. However, in addition to the number of services alone, an inside or outside meter location has a significant impact on the time required to make the connections.

As shown in Figure 2.19, the *connecting services* task accounted for 23% of the manpower necessary to install and maintain a temporary water supply. This percentage holds true whether the meters are inside or outside, but does not necessarily imply that the two variations require the equivalent number of man hours to complete. Figure 3.6 shows the number of man hours applied toward connecting services to the temporary pipelines using data provided by the study group. Trend lines for both inside and outside meter locations are also shown. This data indicates that it takes nearly twice the number of man hours to connect services inside than those outside. This is believed to be largely due to coordinating access with residents for meter or main shut off valve.

Roadway and Driveway Crossings

The physical location of the temporary pipe network is a critical component when assess the necessary installation time. Some of the physical features that must be considered are: the amount of pipe that must be buried either in street or commercial entrance crossings, the number of residential driveways crossed and the method of ramping the pipe for vehicles, wheelchair access crossings.

Street crossings and large commercial driveways typically require excavating enough roadway material to bury the temporary pipe flush with the road surface to protect it from large volumes of traffic (Figures 3.7 and 3.8). Because this effort requires a considerable amount of manpower it is best to minimize road crossings as much as possible. Any such crossing is also a liability to the contractor and the utility. Increased footage of street crossing also makes the water supply more susceptible to interruption due to vehicular damage. A minimal amount of data was collected on temporary pipelines buried in paved areas. The length buried was included as a line item on the field data forms, but manpower required for subtasks (excavation, burial and paving) was not gathered.

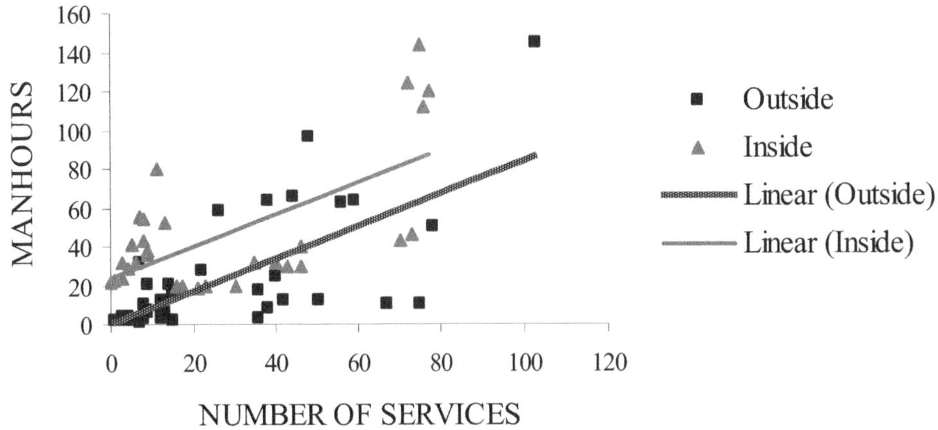

Figure 3.6: Number of man hours to connect services by meter location

Figure 3.7: Fitting used for transition to buried section, temporary asphalt fill protects pipe and general public

Figure 3.8: Nitrile rubber buried flush with pavement prior to temporary asphalt installation

Some temporary pipeline materials such as the nitrile rubber pipe, require no ramps or specialized crossing devices for the smaller diameters (Figure 3.5). PVC, HDPE, and steel pipelines, however, all require some form of ramp to enable customers to cross the pipe without injuring themselves, damaging the pipe, or their vehicles. The 3-inch diameter nitrile rubber hose used as a part of this project does not necessarily need to be ramped, but larger diameters of this material do require some form of vehicular ramp.

As discussed in Chapter 2, there are a number of methods used at driveway crossings that allow residential vehicular traffic to traverse the pipe. The man hours required specifically to ramp residential drives were not gathered, simply the method used. The most commonly employed method was the prefabricated rubber ramp (Figure 3.9). These are quickly installed, offer adequate protection to the pipe, pedestrians, vehicles, are extremely durable, and can be reused. Temporary asphalt, gravel, or a mixture of the two is also an acceptable ramping method (Figure 3.10). Heavy traffic tends to disperse the aggregate and these crossings will require regular maintenance. Asphaltic ramps are beneficial when crossing larger diameter pipe as the ramp slope is easily altered by widening the footprint of the ramp. Wood, less commonly used, is less durable than some other ramping method, but is easily configured to traverse any diameter pipeline (Figure 3.11). Trenching and burying the pipe is likely the most time consuming of all the methods of crossing driveways, (Figure 3.12). This technique is very effective at protecting the pipe and vehicles, but is generally reserved for commercial entrances that experience heavy truck traffic. For additional information on vehicular crossing see Chapter 5, Best Management Practices: Temporary Pipeline Commentary.

Figure 3.9: Prefabricated rubber ramps

Figure 3.10: Temporary asphalt ramp

Figure 3.11: Wooden ramping system

Figure 3.12: Buried line at driveway

Temporary Pipe Manpower Requirements by City

Each participating utility had a unique set of physical, regulatory and operating requirements that contributed to the temporary pipe installation rate. Figure 3.13 shows the amount of labor required by each of the 9-utilities that took part in this study to install 1,000-feet of temporary pipeline. The differences in production rates depend on many factors including temporary pipe material, temporary pipe diameter, length of pipe buried in pavement, and number of driveways crossed and ramping method.

Of the 9-utilities surveyed, NKWD installed temporary pipe with the greatest efficiency at a rate of 3.3- man hours per 1,000-feet. AACDPW rated the second most efficient installing temporary lines at 5.2-man hours for every 1,000-feet. Baltimore, Charlotte, Cleveland, Louisville, Nashville, and Omaha, averaged between 14.3 and 28.2 man hours per 1,000-feet. On the lining project in Cincinnati, the crew averaged a manpower utilization rate of 146.2-man hours/1,000-linear feet of temporary pipeline installed.

The highly varying rate for temporary pipe line installation is generally attributed to any one, or combination of factors. Based on the data, it is believed that NKWD was highly efficient at installing temporary pipe due to the use of the 3-inch nitrile pipe material. The nitrile material was highly flexible, easily installed and did not require extensive ramping or protection. Conversely, the data reveals that the same task required GCWW a disproportionately larger level of manpower to complete. GCWW's elevated manpower utilization is likely due to a combination of issues including temporary pipe diameter and ramping materials. GCWW exclusively used 4-inch diameter PVC pipe for the entire temporary water supply distribution system installed as part of this study. As shown in Figure 3.1, installation of 4-inch PVC requires twice the manpower of 2-inch PVC or 2-inch HDPE. In addition to the 4-inch temporary pipe, GCWW used asphalt ramps at 75 percent of the driveway crossing locations as compared to 22 percent within the study group (Figure 2.23). While not specifically studied, it is believed that the asphalt ramps require more manpower to install then the reusable rubber ramping systems.

Temporary Pipe Material Durability

Recurring leaks and breaks on temporary lines can be very costly to the contractor in terms of manpower as well as costly to the utility in terms of customer confidence. The various materials have varying break frequencies, in part due to the material itself, but also due in part to the site conditions. The number of breaks was logged for every segment of temporary line and analyzed by material type, number of days in service, and number of driveways crossed.

Break Frequency by Pipe Material

Figure 3.14 is an illustration of break frequency per 1,000 linear feet of temporary pipeline installed, by material type. Based on the data collected, HDPE outperforms the other two materials with an average rate of 0.40-breaks/1,000lf. PVC materials exhibit the greatest number of breaks with a rate of 1.15 breaks/1,000lf. PVC materials were used by 7 of the utilities involved in the study and accounted for approximately 59% of the total temporary pipe footage installed. The number of days the temporary pipe was in service is not included in this portion of the analysis.

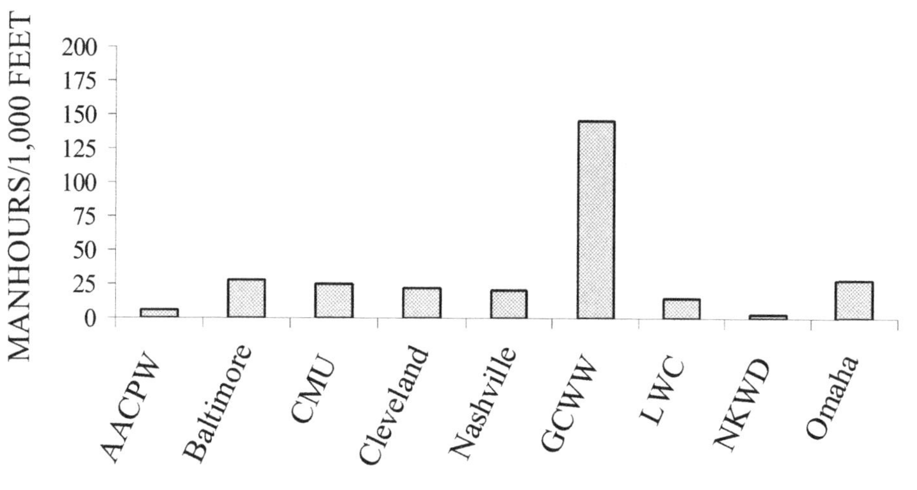

Figure 3.13: Average man hours/1,000 feet of temporary pipe installed

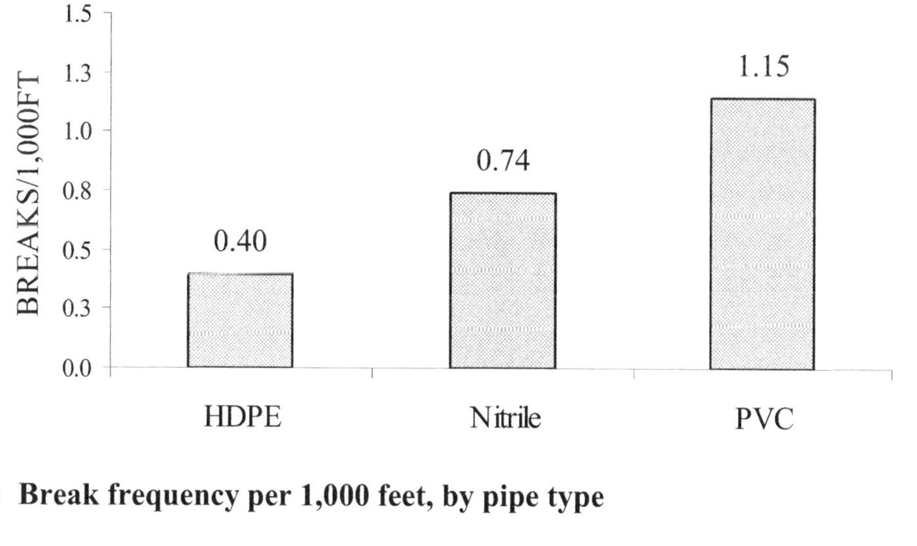

Figure 3.14: Break frequency per 1,000 feet, by pipe type

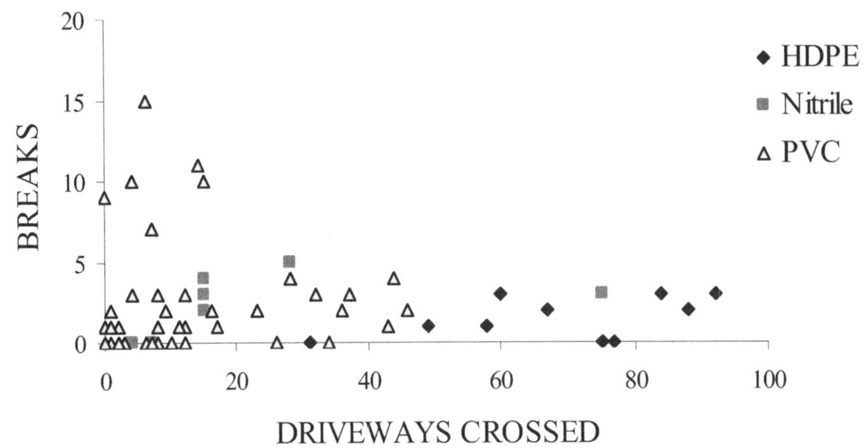

Figure 3.15: Number of driveways crossed vs. number of breaks, by pipe type

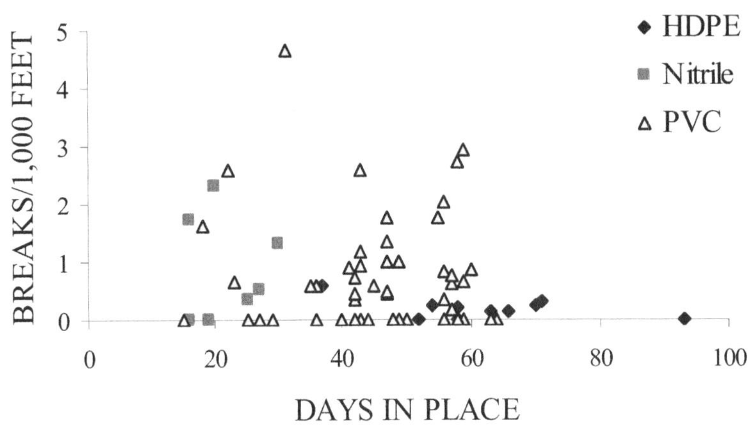

Figure 3.16: Number of days temporary pipe was in place vs. number of breaks per 1,000-feet, shown by pipe type

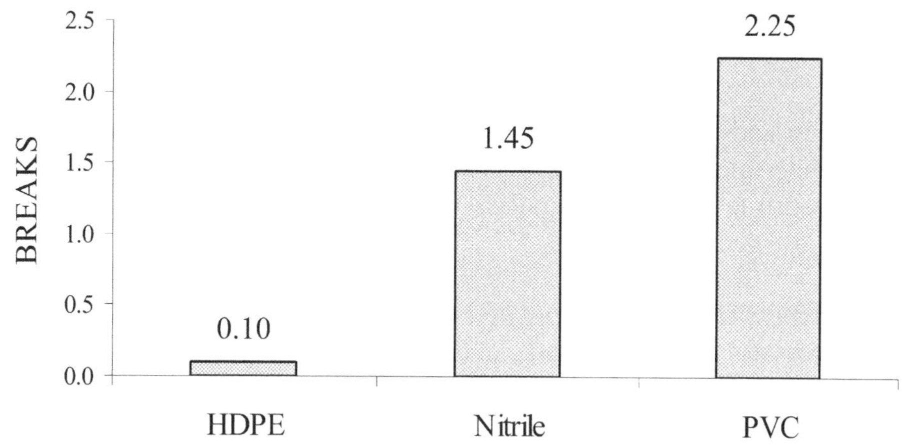

Figure 3.17: Average number of breaks per 10-driveway crossings

Number of Breaks by Driveway

The durability of the temporary pipe system is a critical item for the design of any temporary supply network. Often, the most critical area of the temporary pipe network is the customer driveway crossing as it is highly visible, mandates significant customer interaction, and is subjected to concentrated stress. Thus, driveway breaks are not only time consuming and expensive to repair, but also the source of customer complaints.

The temporary pipe durability was assessed with respect to both the number of driveways crossed and the length of time in service. It was anticipated that as both factors increased, the number of breaks would also increase. The data were further subdivided into material types to determine the role of material selection on pipe breaks. The data plotted in Figures 3.15 and 3.16 display the expected break trends for both number of driveways and length in service. Additionally, the HDPE materials appear to perform well at driveway locations and have a substantially lower break rate.

Based on the data obtaincd, it is obvious that a variety of variables affect the temporary pipe break rate. To concurrently account for these variables, the data was further analyzed to normalize it for length, time and number of driveways crossed. On average each 500-foot lining segment (1,000 feet of temporary pipe) encountered ten driveway crossings and remained in service for 26-days. Analyzing the data in this manner shows HDPE outperformed both PVC and Nitrile (Figure 3.17).

DETAILED DATA ASSESSMENT

One of the objectives of this research was to identify cleaning and lining programs operated efficiently. However, there are a variety of individual tasks associated with the cleaning and lining process, and some utilities excelled in specific areas while falling short in others. To facilitate the assessment, the cleaning and lining process was subdivided into specific task items and utilities recorded the time and man hours necessary to complete each task. It was important to record both days and man hours so that contractor performance and the customer's perception of performance could be measured. In this manner, the length of time, or effort, required to complete each task for both average and efficient operations could be identified.

Task Definitions

The cleaning and lining process was subdivided into separate work items to facilitate critical review of the operations. Below is the task sequence a pipe lining project normally progresses. Pre-construction work activities were not included within this study as it is not directly part of the defined "downtime".

- *Install temporary water supply pipe*: Unload, layout, and join temporary lines. Make the necessary street, sidewalk, and driveway cuts to bury pipe per specifications. Install approved ramps at wheelchair crossings and residential driveways where appropriate.
- *Disinfect temporary water supply pipe*: Assemble chlorination equipment and make necessary connections to the temporary pipeline. Fill pipe at a defined rate to assure specified chlorine concentration, releasing air at any high points.
- *Flush temporary water supply pipe*: Assemble de-chlorination equipment and make necessary connections to the temporary pipeline. Flush pipe at a rate that assures the chlorine is neutralized. Disconnect equipment when de-chlorination is complete and allow pipeline to flush.
- *Sample temporary water supply pipe*: Collect samples for bacteriological analysis in sterile bottles in accordance with AWWA standard.
- *Approve temporary water supply pipe:* verify samples pass bacteriological testing.
- *Connect services to temporary water supply:* Establish the necessary connections between the temporary pipeline and the customers' services, flushing both the customers' point of connection and temporary hose prior to connecting. These connections may be at exterior meter vaults, hose bib connections at the house, or in the basements/crawlspace.
- *Dewater existing pipe*: operating valves to isolate pipe section to be lined, cut and remove segments of main in access pits, pump drained water to waste
- *Clean existing pipe*: Remove corrosion from inside existing pipe using a drag scraper, rack feed bore, or other cleaning method.
- *Line existing pipe:* Cement Mortar - set up lining equipment, mix batches of cement mortar, pull lining hose to remote pipe access point of pipe to be lined, pump initial cement mortar mixture into empty hose, start lining machine and verify proper consistency of cement mortar, winch lining machine towards initial access point at a constant rate and pump pressure to achieve specified lining thickness, clean up equipment following lining; Epoxy Coating – circulate epoxy components through holding tanks and hoses heating them to the appropriate temperature, perform weight checks on individual components, pull lining hoses or umbilical line to remote access point of pipe to be lined, start lining machine and verify proper component mixture, winch lining machine to initial access point at required rate and pump pressure to achieve specified coating thickness, clean equipment following lining.
- *Provide time for pipe to cure:* Allow pipe lining materials to cure and inspect pipe interior for satisfactory application.

- *Disinfect existing pipe*: Fill newly lined pipe with hyper chlorinated water (see chlorinating temporary pipeline)
- *Flush existing pipe:* (similar to temporary pipeline above)
- *Sample existing pipe*: (similar to temporary pipeline above)
- *Approve existing pipe:* verify samples pass bacteriological testing.
- *Reinstate existing pipe*: Following approval, open isolation valves individually to flush each intersecting main before opening newly line pipe to the system.
- *Disconnecting temporary supply:* Disassemble temporary pipeline_from customers' service lines, back flush lines, reinstall meters, connect customer services.
- *Remove temporary water pipe*: Remove disassembled temporary pipeline and appurtenances from construction site, restore paving and landscape as necessary.

Problems in Analyzing Data Directly

As mentioned previously, the data collected for this study originated from nine separate utilities and 74 separate lining segments. Each project was slightly different in terms of length, equipment, crew, etc., and direct comparison of the data would not provide meaningful results. As an example, a calculated average value of pipe cleaning time would be inappropriately skewed if the data set contained one very long run.

Preliminary assessment of the data indicated that run length was a controlling variable. As such, the data were normalized to a standard unit of one city block to enable direct comparison. The standard one city block unit was assumed to include 500 feet of lined pipe and 1,000 feet of temporary water supply. The 500 foot length was selected as it is a standard block length in old city neighborhoods; areas that are likely to have old cast iron water distribution systems. The 2:1 ratio of temporary water supply to lined pipe was based on Figure 2.15 that established the ratio based on collected data.

In some instances, the data collected did not adequately account for the calendar duration of a task. While man hour data were directly recorded, calendar duration of tasks were determined by subtracting the date the previous task was completed from the date the task of interest was completed. This method worked when a task spanned at least one day, however, if multiple tasks were completed within a day, a zero value was calculated. To avoid biasing the data with zero values, the zero values were considered null values in the data analysis.

The Need for Statistics

When large amounts of data are reviewed, statistics can provide an effective means to succinctly characterize the trends. "Averaging" data is likely the most commonly used statistical technique; however, in standard statistics there are three methods to determine an arithmetic average. Each method may more or less appropriately represent the data, thus, a clear understanding of each is necessary. Within standard statistics, the three methods to measure central tendency are defined as:

- *Mean:* sum of the data divided by the number of data pieces
- *Mode:* value of the most frequently occurring data piece
- *Median:* the middle number of the data set when the data pieces are arranged into ascending or descending order

In addition to the average standard deviation is also commonly used to define large data sets. The standard deviation represents how tightly the data is clustered around the mean value. A low standard deviation value indicates the data values are all very close to the mean value and a large value indicates the data are wide ranging. If the data is normally distributed, then half of the data pieces within a set will be captured between plus-one and minus-one standard deviation increment of the mean

An example of standard statistics applied to the cleaning data collected is shown in Figure 3.18. The mean average to clean only a 500 ft length of pipe was 4.36 days with a large 4.9 standard deviation. While these values technically represent the data, they do not represent the data trends. The reason the trends are not adequately represented is that the standard statistics values are skewed by relatively few very large values.

A more effective way to represent the data collected is to use the median value and the inter quartile range. Similar to standard deviation, the inter quartile range (IQR) measures how tightly the data is clustered around the *median* value. The IQR is the minimum range required to capture half of the data. A low IQR value indicates the data values are all very close to the median value and a large value indicates the data are widely ranging. For the cleaning data collected, the median length of time required to clean pipe was 2.15 days and the IQR was 4.97. The IQR value represents a much tighter data clustering than the standard deviation (half of the data is captured within two standard deviation units, 9.8).

To visualize the median and IQR concept a box plot was developed to graphically depict the values and other important data information. Figure 3.19 displays the cleaning information as depicted in Table 3.1 where the shaded region represents the IQR of 4.97 days. The horizontal line inside of the shaded rectangle represents the median of 2.15 days. The mean is represented as a circled addition sign at 2.2 days. The vertical line protruding up and down from the IQR region are "whiskers" that represent the data that fall within a 1.5 multiplier of the IQR on either side of the box. A data point far from the typical value and outside the whisker range is considered an "outlier" and represented by an asterisk.

Results of Statistical Analysis

The box plot statistical analysis procedure was used to assess the cleaning and lining task data collected. The tasks were divided into two groups representing activities associated with the temporary water supply and activities associated with the cleaning and lining operation. This data was assessed based on man hours and calendar days necessary to complete the tasks. Summaries of the collected data are presented in Tables 3.1 to 3.4 and Figures 3.20 to 3.23.

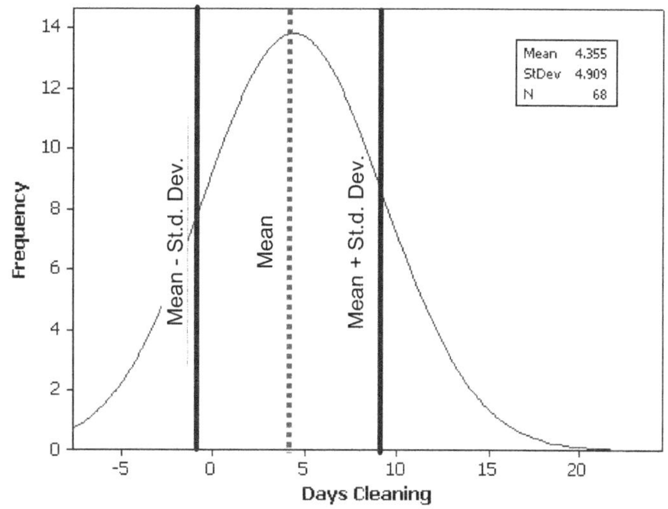

Figure 3.18: Normal distribution curve for days required to clean pipe, normalized to 500 ft.

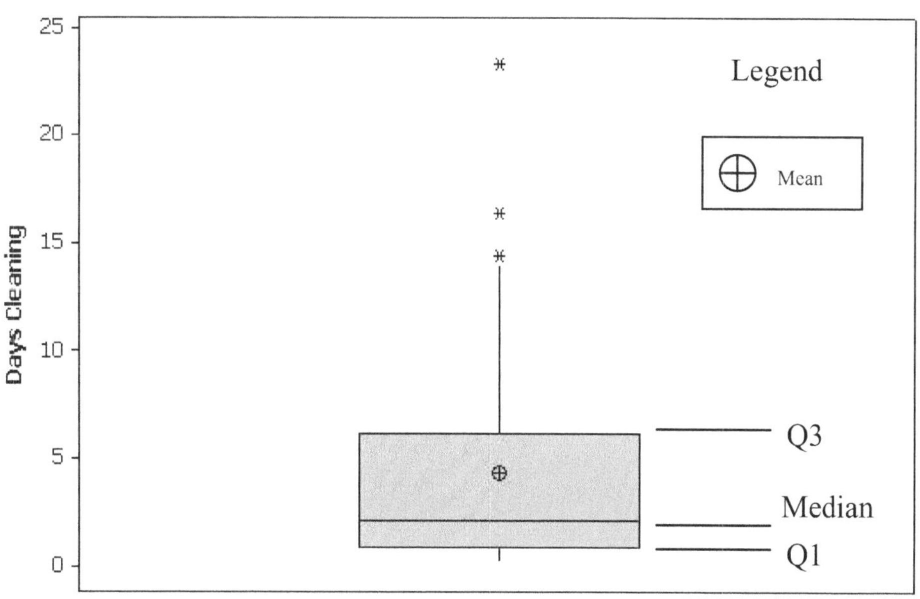

Figure 3.19: Box plot of calendar days required to clean existing pipe, normalized to 500 ft.

Table 3.1
Statistical summary of calendar days required to complete cleaning and lining tasks, normalized to 500 ft

Statistical Value	Dewater (days)	Clean (days)	Line (days)	Disinfect (days)	Flush (days)	Sample days)	Approve (days)	Reinstate (days)
Mean	2.5	4.4	6.7	2.0	2.0	0.9	1.6	4.6
Low Value	0.3	0.2	0.4	0.0	0.2	0.2	0.2	0.3
1st Quartile	0.9	0.9	2.6	0.3	0.7	0.3	0.5	0.7
Median	**1.7**	**2.2**	**5.9**	**1.0**	**1.1**	**0.8**	**0.9**	**1.1**
3rd Quartile	3.6	5.9	14.9	2.1	1.7	1.3	1.7	2.0
High Value	10.0	23.3	51.4	13.4	15.0	5.0	10.6	48.6

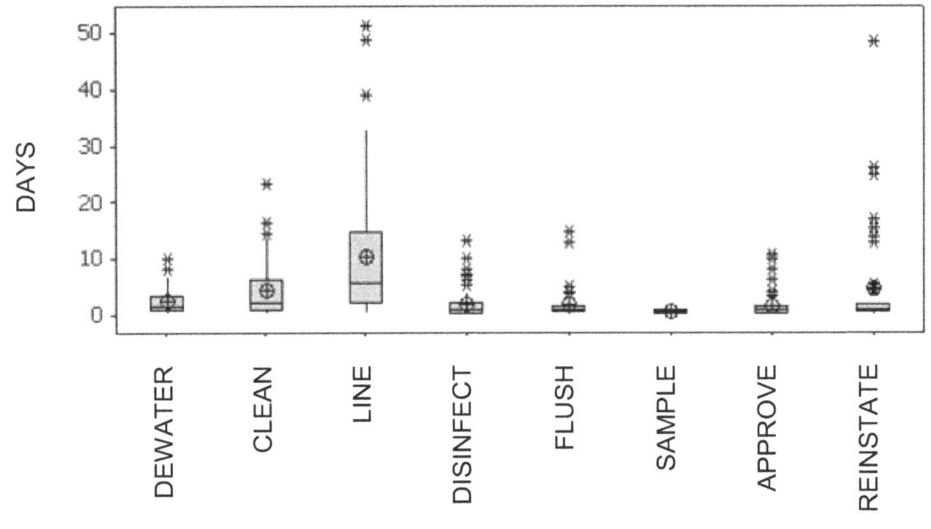

Figure 3.20: Box plot of days required to complete cleaning and lining tasks, normalized to 500 ft.

Table 3.2
Statistical summary of man hours required to complete cleaning and lining tasks, normalized to 500 ft

Statistical Value	Dewater (man hours)	Clean (man hours)	Line (man hours)	Disinfect (man hours)	Flush (man hours)	Sample (man hours)	Approve (man hours)	Reinstate (man hours)
Mean	6.8	19.8	23.0	18.7	6.1	3.5	7.6	65.8
Low Value	0.7	2.1	1.3	0.0	0.3	0.2	0.2	0.2
1st Quartile	1.8	10.1	11.1	0.9	0.9	0.6	0.5	1.0
Median	**4.1**	**17.5**	**18.4**	**1.4**	**2.2**	**1.4**	**1.3**	**22.8**
3rd Quartile	6.9	33.0	31.7	4.2	4.3	3.2	7.1	109.2
High Value	45.7	60.0	60.0	72.1	154.8	16.4	154.8	440.0

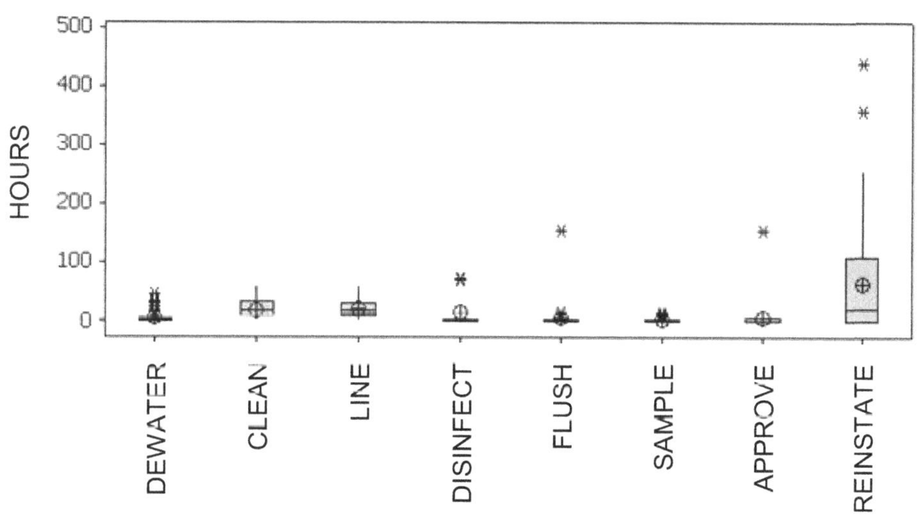

Figure 3.21: Box plot of man hours required to complete cleaning and lining tasks, normalized to 500 ft

Table 3.3
Statistical summary of calendar days required to complete temporary water supply tasks, normalized to 1,000 ft (associated with 500 ft of pipe renewal)

Statistical Value	Install (days)	Disinfect (days)	Flush (days)	Sample (days)	Approve (days)	Connect Services (days)	Days in Service (days)	Days on Ground (days)
Mean	6.0	2.7	1.7	1.1	2.0	5.9	33.0	48.2
Low Value	0.3	0.0	0.1	0.1	0.2	0.2	2.9	4.2
1st Quartile	1.0	0.3	0.6	0.3	0.5	1.5	8.2	14.1
Median	**2.8**	**0.8**	**1.0**	**0.9**	**1.1**	**3.0**	**17.8**	**26.1**
3rd Quartile	7.8	2.8	2.2	1.6	2.4	6.6	44.0	65.2
High Value	29.0	20.0	6.7	3.3	10.0	74.7	173.3	206.5

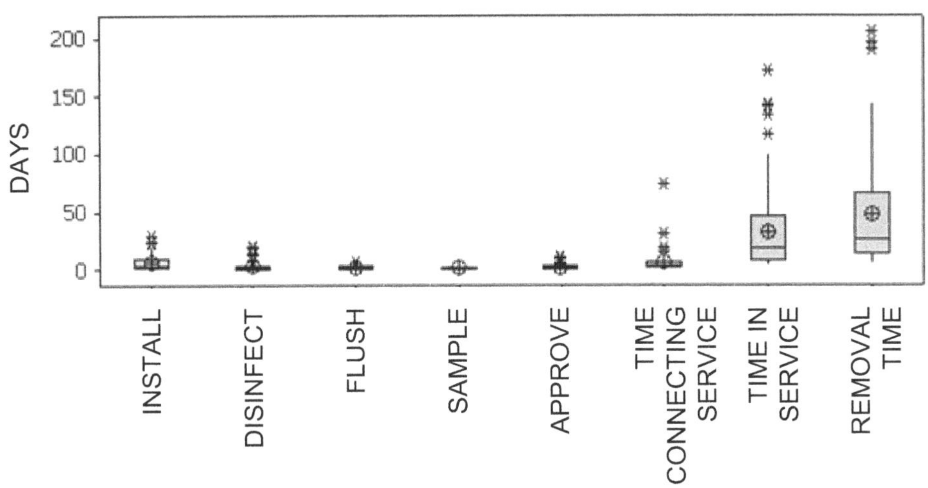

Figure 3.22: Box plot of calendar days required to complete temporary water supply tasks, normalized to 1,000 ft

Table 3.4
Statistical summary of man hours required to complete temporary water supply tasks, normalized to 1,000 ft

Statistical Value	Install (man hours)	Disinfect (man hours)	Flush (man hours)	Sample (man hours)	Aprove (man hours)	Connect (man hours)	Remove Services (man hours)
Mean	42.0	35.7	6.2	4.6	10.0	23.5	13.2
Low Value	1.8	0.0	0.2	0.1	0.1	0.5	0.9
1st Quartile	8.9	0.4	0.8	0.7	0.6	4.8	6.5
Median	**21.8**	**1.1**	**2.1**	**1.4**	**1.9**	**12.0**	**11.4**
3rd Quartile	35.1	3.3	7.3	4.4	13.3	30.8	18.2
High Value	146.3	143.3	80.0	26.7	160.0	72.1	60.0

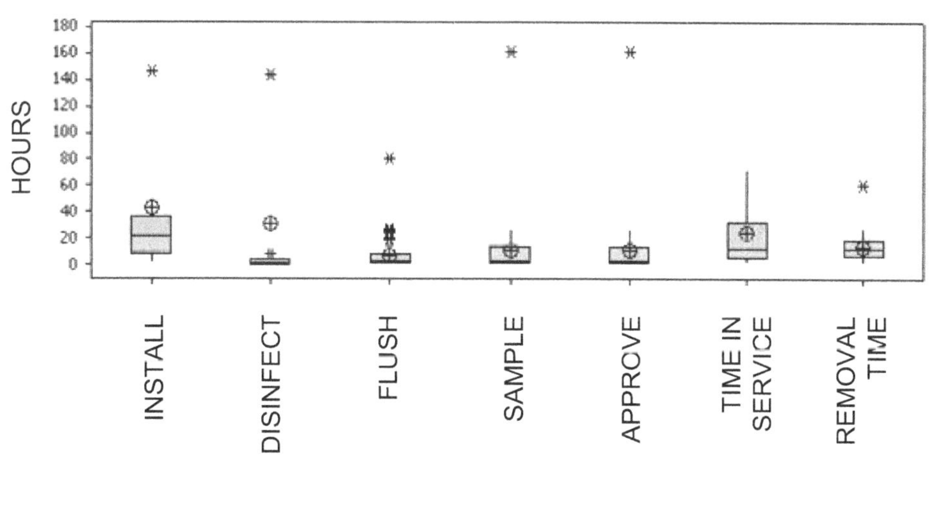

Figure 3.23: Box Plot of man hours required to complete temporary water supply tasks, normalized to 1,000 ft

CUSTOMER COMPLAINTS

The number of complaints registered by residents on lining projects was of primary interest to the study group. A break in the temporary line can create water outage, or very low pressure, for customers with no prior notification whatsoever. The data collected indicated that about 9% of the customers registered some form of complaint during the pipe renewal projects within this study. Data from documented customer complaints were charted first as a function of the number of breaks that occurred on each pipe segment; second as they related to the number of days the temporary pipeline was in place, and third as a function of the number of services connected to the temporary pipeline.

The relationship between the number of temporary line breaks and the total number of complaints is shown in Figure 3.24. The break/complaint date was plotted as it was collected from the participants without any normalizing for various segment lengths, material type, etc. As anticipated, the number of complaints correlates to the number of temporary pipe breaks.

In addition to break frequency, it is anticipated that the number of customer complaints would increase with respect to the number of days the temporary pipeline is in place. The temporary pipe is perceived as an inconvenience and it is believed that residents will quickly tire of the nuisance and the construction activity. Additionally every extra day the temporary water supply pipeline and service hoses are in place presents an opportunity for the pipe to break, a pedestrian to trip, to damage a vehicle or damage personal property.

Figure 3.25 displays the number of complaints with respect to time the temporary service pipe was installed. The "days in place" were calculated by using the difference between the day the temporary pipe installation began and the day it was removed from the site. Additionally, to account for varying segment lengths, the complaints were normalized per 1,000 feet of temporary pipe installed. No distinction was made for varying pipe materials.

Based on the data collected, there does not appear to be a correlation between the length of time the temporary pipe is in place and the number of complaints. Of the 74-segments of temporary pipe installed 73-included break data. Of those 73-segments 38% reported zero breaks and 61% had 1-break or less. Thus, it is likely that complaints are more closely correlated with breaks than project duration.

The service connections create a number of potential customer complaint opportunities. Figure 3.26 shows the increasing relationship between the numbers of complaints and the number of services per pipe segment. Common customer complaints include:

- Not wanting water outages to occur while they are home
- Difficulty of coordinating with contractor to provide access to in-house meters and valves
- Noise and mess
- Contractor personnel interaction
- Line breaks
- Equipment/service hazards

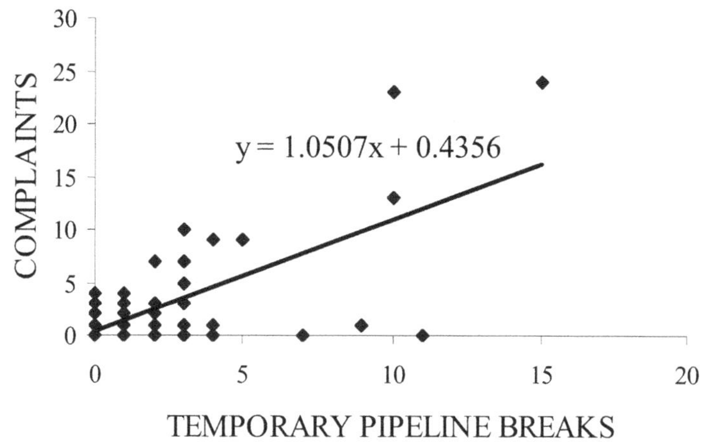

Figure 3.24: Temporary pipeline breaks effect on customer complaints

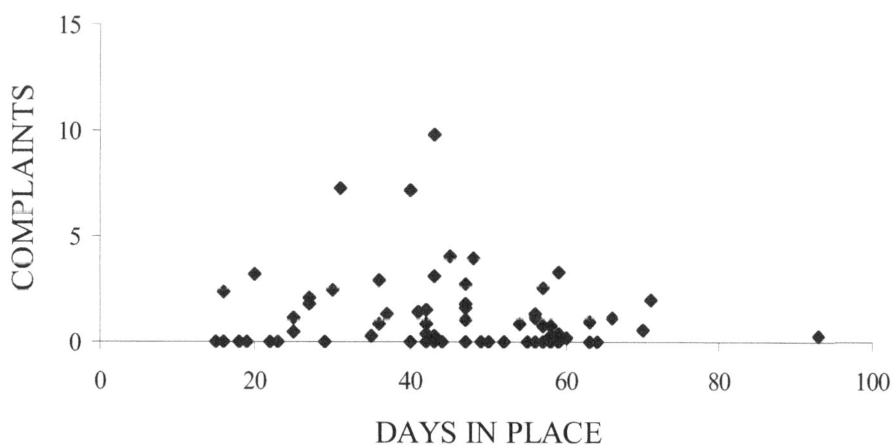

Figure 3.25: Complaints per 1,000 feet of temporary pipe, shown in number of days

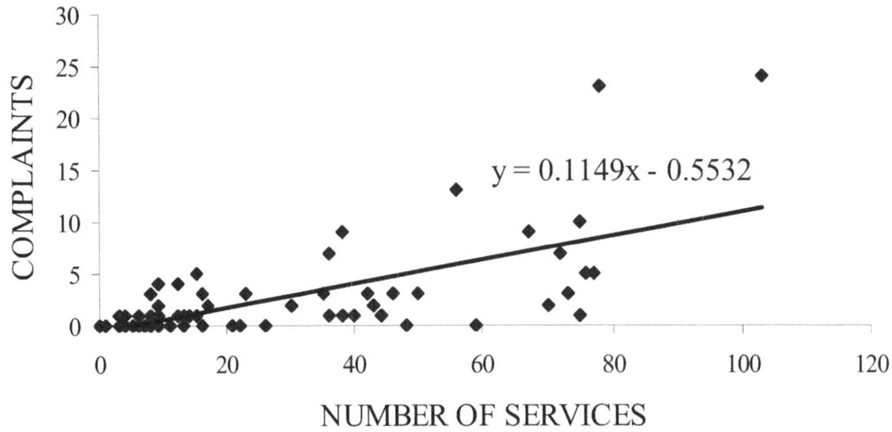

Figure 3.26: Number of complaints as a function of number of service attachments

CHAPTER 4
UTILITY PARTICIPANT CASE STUDIES

INTRODUCTION

Nine utilities provided lining data in support of this project. Listed below is a brief description of the general processes and characteristics of each system. The task rate information collected for each utility has then been compared to average rates determined from all utility data. With this information, utility performance with respect to lining and temporary pipe operations can be assessed to typical values for both man hours and days of service.

BALTIMORE METROPOLITAN WATER DISTRICT

Program Background

The Baltimore Metropolitan Water Distribution System supplies potable water to Baltimore City, Baltimore County, Howard, and Anne Arundel Counties. In addition, raw water is sold to the Counties of Harford and Carroll. The water system serves 1.8 million people an average of 241 million gallons of water per day through three separate water treatment plants at the time when this case study was completed. A fourth water treatment plant will be coming on-line in the future.

The water transmission and distribution system contains approximately 4,100-miles of pipeline over a 291-square mile distribution area. The system includes 65,072 valves (4- to 108-inch) and 21,422 fire hydrants. Baltimore began installing cast iron in the mid 1800's and continued to install unlined materials until about 1956. After 1956 Baltimore has primarily been purchasing and installing factory lined ductile iron pipe. Within Baltimorc's system, approximately 80% (3,000 miles) of the water mains are unlined ductile iron and 15% are factory lined cast iron.

Water main cleaning and lining has been a routine part of Baltimore's capital improvement program since the 1960s. The cleaning and lining program was initiated to improve low flows, reduce discolored water complaints, and increase fire protection. As such, most of the lining work has focused on the distribution system. However, some transmissions mains have been included in the renewal effort.

Typical Project Construction

The intent of the Baltimore Metropolitan Water Distribution System infrastructure renewal program is to systematically upgrade the existing water distribution system to current standards. This work involves the cleaning and lining of existing water mains, some new main installations, service renewals and replacements and installation of meter settings. All valves, 20-inch and smaller and all fire hydrants are replaced, all the "old type" remote read services are replaced with new outside metered services and all galvanized service lines are replaced with copper piping. Generally dead end mains are connected, all lead joint tees below 20-inch are replaced and anything that needs to be corrected is addressed. During this case study the common practice was to install a new metered service outside, to replace the "old type" remote read services; however Baltimore is now entertaining the idea of replacing the "old type" remote read services with a new type of "ride by" remote read meters, inside the house.

During the lining operations, a temporary water system is installed above ground to provide domestic service and fire protection. The temporary piping is made of either steel, plastic, galvanized iron pipe or hose which are connected to consumers in a pre-determined area. Temporary fire hydrants and valves are installed and the piping is chlorinated, flushed and tested; similar to the installation procedures of a new water main.

Once the temporary water system is established the isolated lines are cleaned and lined. To clean the lines a rod is inserted into the pipe with a "rodding" truck, which inserts a cable through the pipeline. A "pig" is inserted into the pipe and pulled through the pipeline by the cable attached to a wench truck. This process is repeated with increasing pig diameters until the built up corrosion material (tuberculation) is removed. After the pipe is cleaned a reinforced cement mortar hose is attached to the cable, and a "spun lining machine", is pulled through the pipeline at a specified rate of speed which sprays between 3/16" to ¼" of cement mortar onto the pipe walls and also trowels the mortar smooth. The cement mortar lining thickness is checked at the ends of the open pipe and then covered with plastic to protect the interior and slow the curing time of the cement lining. Once the cement has cured, hyper-chlorinated water is introduced into the main for a 24-hour contact time before sampling. After sampling, testing, and approving the newly lined pipes are returned to service and the temporary lines are removed.

The data collected in the above referenced 5,245 linear feet of rehabilitated pipe includes three runs of undefined lengths of cleaning and lining with an indeterminate number of replaced remote read services which would inflate all four of the following data charts slightly.

Study Area Description

The data contained in this report was collected residential lining project performed in autumn 2004. Data was obtained from three separate cleaning and lining runs for a total 5,245 linear feet of lining. Pipe diameters of the rehabilitated mains were 6-inch and 8-inch. During the rehabilitation, water service was maintained to 237-service connections by placing 11,950 linear feet of 2-inch and 4-inch diameter PVC and steel temporary pipe.

Task Rate Comparison

Figures 4.1 through 4.4 display Baltimore's pipe rehabilitation program as compared to other utilities participating in the study. The data are segregated by lining and temporary service tasks and also by man hours and days per task. The solid line running horizontally across the figure represents Baltimore's data. The background box plot represents the average utility data as analyzed and presented in Chapter 3.

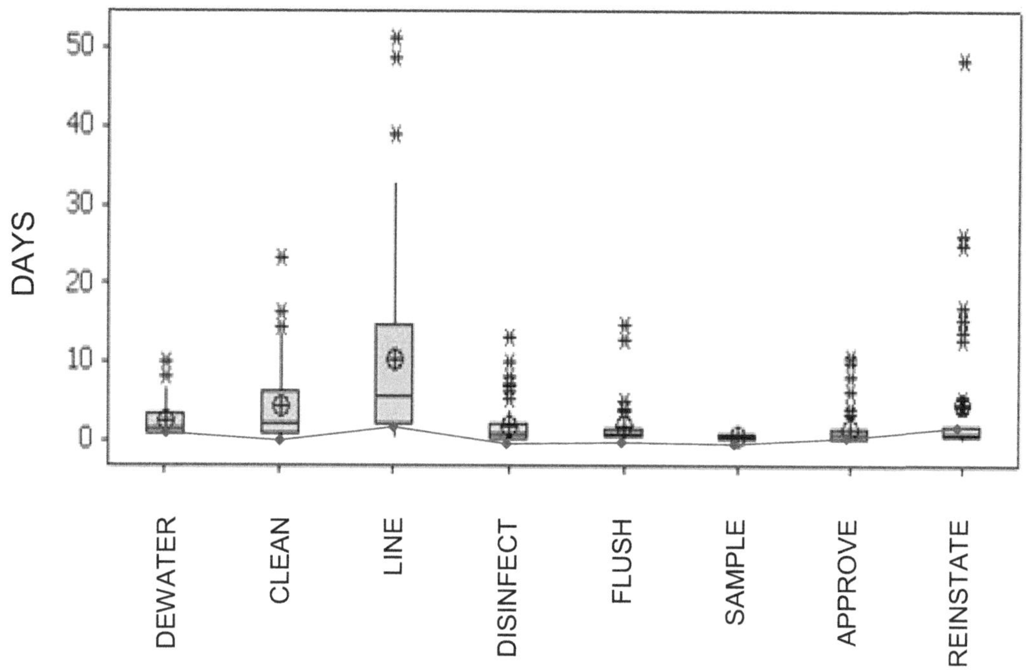

Figure 4.1: Comparison of Baltimore's cleaning and lining operation by work day

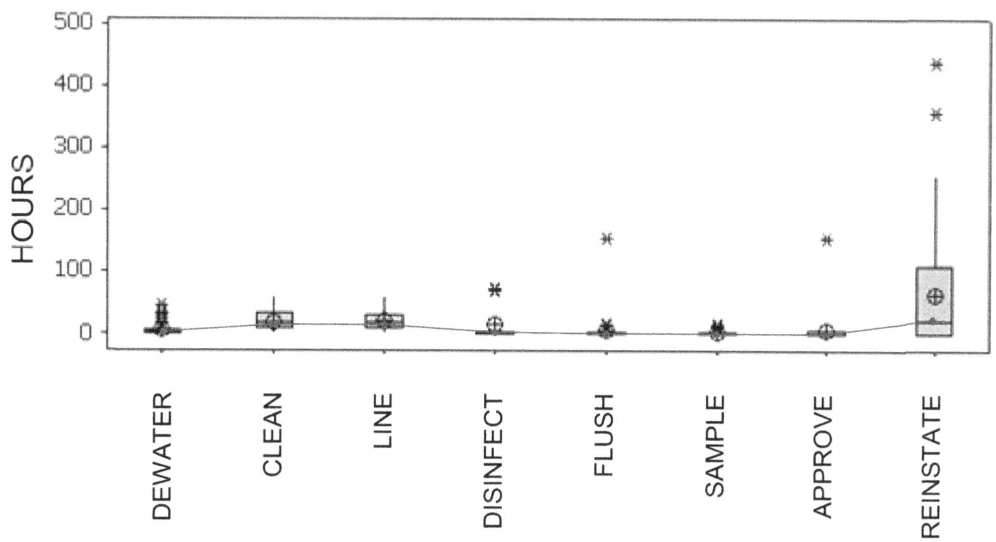

Figure 4.2: Comparison of Baltimore's cleaning and lining operation by man hour

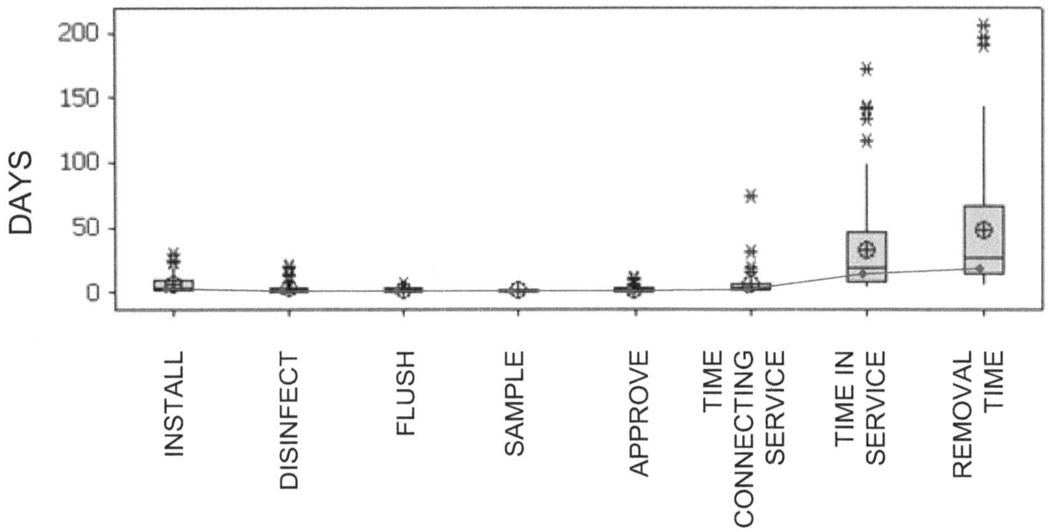

Figure 4.3: Comparison of Baltimore's temporary water supply operation by work day

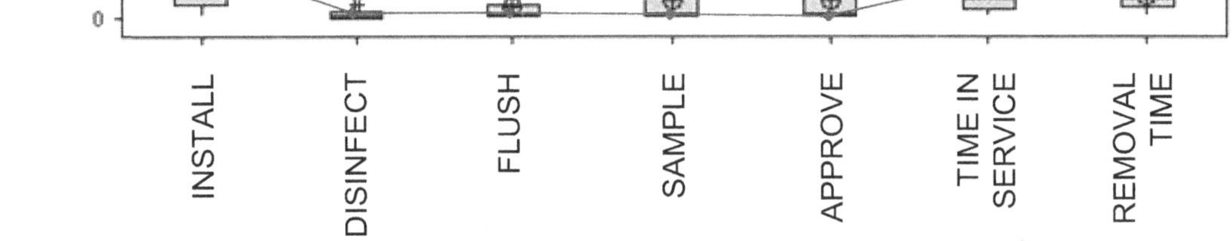

Figure 4.4: Comparison of Baltimore's temporary water supply operation by man hour

NORTHERN KENTUCKY WATER DISTRICT

Program Background

Northern Kentucky Water District (NKWD) is the largest geographic water district in Kentucky and provides drinking water to over 76,000 customers in Kenton and Campbell counties. NKWD operates and maintains approximately 1,030-miles of water main in its transmission and distribution pipeline network. Pipe sizes range from 3-inches in diameter to 42-inches in diameter.

Construction on NKWD's supply system started in 1889 and the initial 30-inch diameter line remains in service today. Between 1889 and 1960 NKWD purchased and installed unlined cast iron pipe for system expansion. Since 1960 the utility has primarily installed cement mortar lined cast iron and cement mortar lined ductile iron mains for corrosion protection. Within NKWD's system approximately 267-miles (26%) is unlined cast iron with 3- to 30-inches pipe diameters.

NKWD's first in situ lining took place in 1986 with the cement mortar lining of a 30-inch diameter transmission main. In 1999, driven by the need to improve fire flow and reduce discolored water complaints NKWD embarked on an annual cleaning and epoxy lining program. Currently, NKWD epoxy lines 2.5- to 3-miles of unlined cast iron pipe each year.

Customer Relations

The Northern Kentucky Water District (NKWD) believes that communication between both the contractor and the customer is a key factor to the success of its pipe renewal projects. To facilitate the lining process, NKWD representatives review the Contractor's project schedule before project start-up and identify key performance criteria. Close coordination is then maintained throughout the project to ensure satisfactory performance.

Within the community, NKWD initiates an aggressive communication plan to inform customers of planned work activities. Business and commercial customers are contacted several weeks in advance of the scheduled work to account for the complexities associated with water interruptions to their processes. At a minimum of one week prior to construction, a notice is sent to all affected residential customers. The notices include: a list of streets affected, anticipated duration, the name of the contractor, a description of the work, a caution for customers to inspect their water prior to using it for washing, a statement about restoration of their property, the name of the NKWD Construction Inspector assigned to the project, and contact phone numbers in case customers have questions or emergencies arise.

An equally important aspect of NKWD's communication requirements is to maintain constant contact with the local fire fighting units and Public Works Departments. The construction inspector is to inform the fire department which fire hydrants will be out of commission during the project. The fire department is also informed which fire hydrants are to be used for temporary water supply. Provisions are made, that in the event of an emergency, the temporary waterlines can easily be disconnected from the hydrant leaving the hydrant unencumbered to be used for fire fighting purposes. The benefit of maintaining Public Works Department communication is that the work can be coordinated with other planned projects so as to reduce the inconvenience to area residents.

Typical Project Construction

NKWD has embarked on a multi-year epoxy lining program to improve fire flow and reduce discolored water complaints. The majority of lines rehabilitated are 6- to 12-inch mains within residential neighborhoods. Lining runs are generally limited 500 feet even due to the spacing of the existing fire hydrants and the hilly terrain of Northern Kentucky.

During the lining process, NKWD gives a high level of importance to maintaining an uninterrupted water supply to all customers. To achieve this goal, a temporary water supply is provided during the lining process. The temporary system consists of a flexible rubber hose laid above ground in the street gutter pan or in the utility strip adjacent to both sides of the street. For protection, rubber, steel, or asphalt cold patch ramps may be installed at driveway crossings. Ramps may also be installed at sidewalk crossings to assist pedestrians in negotiating the temporary mains. Once installed, the temporary main is disinfected and flushed. A sample of the water is tested at NKWD's laboratory in order to verify it meets State water quality standards. With laboratory approval, service lines are connected from the temporary waterline to each customer's meter setting to provide continued water service throughout the project. The connection of the water service line, to the temporary main, is made using a flexible 1-inch diameter pipe connected to the meter setting near the street and typically involves less than 10-minutes of water service interruption.

Once the temporary pipe network is established, service pits are excavated and the pipe is cleaned of corrosion using a rack feed boring machine. The rack feed or power bore is preferred for the NKWD system as it does not put undue stress on the existing cast iron mains and has consistently provided quality results. After the cleaning a specialized closed circuit television (CCTV) camera is inserted into the main to verify the effectiveness of the cleaning, integrity of the pipe, and to identify any unknown valves or tees in the line which may need to be removed.

NKWD specifies an epoxy lining product conforming to ANSI/NSF Standard 61. The spray epoxy is applied within 45-minutes per 500 linear foot section and allowed to cure for 16 hours. Following cure, lining quality is inspected using CCTV. Upon new lining approval, the main is disinfected and tested for water quality. Once approved the newly lined main is returned to service, the customers' service lines are reinstated, temporary pipelines removed.

Study Area Description

Data for this report analysis originated from a residential area with an unlined cast iron gridded network and relatively few dead ends. Data were obtained from seven separate cleaning and lining segments for a total 10,850 linear feet of lining. Rehabilitated pipe diameters ranged from 4- to 10-inches. During the rehabilitation, water service was maintained to 245-service connections by 21,900 linear feet of 3-inch diameter nitrile rubber temporary pipe.

Task Rate Comparison

Figures 4.5 through 4.8 display Northern Kentucky Water District's pipe rehabilitation program as compared to other utilities participating in the study. The data are segregated by lining and temporary service tasks and also by man hours and days per task. The solid line running horizontally across the figure, represents NKWD data. The background box plot represents the average utility data as analyzed and presented in Chapter 3.

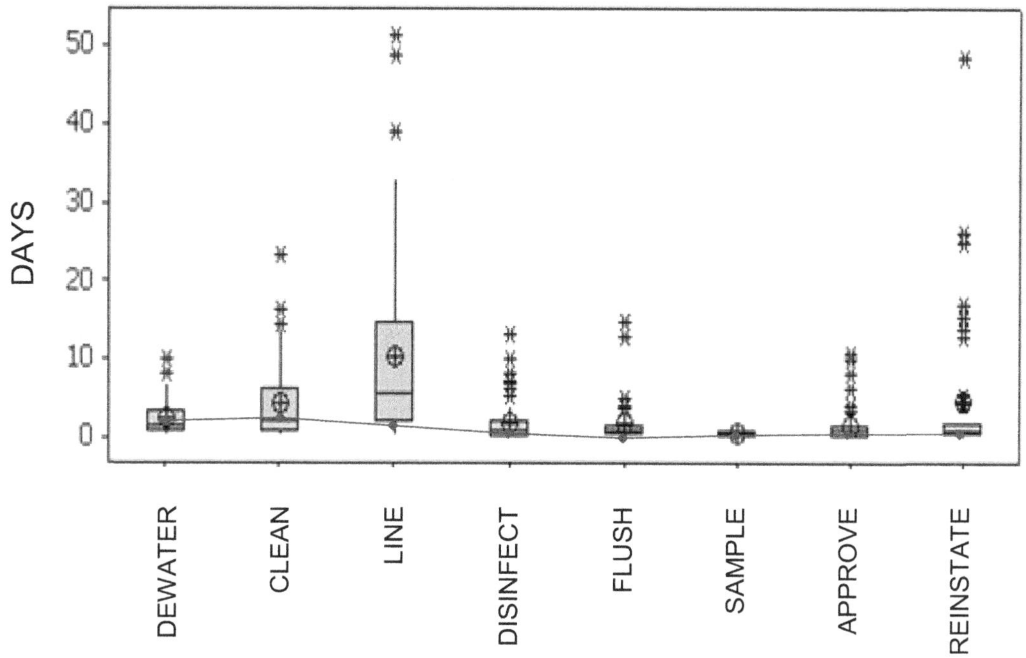

Figure 4.5: Comparison of Northern Kentucky Water District's cleaning and lining operation by work day

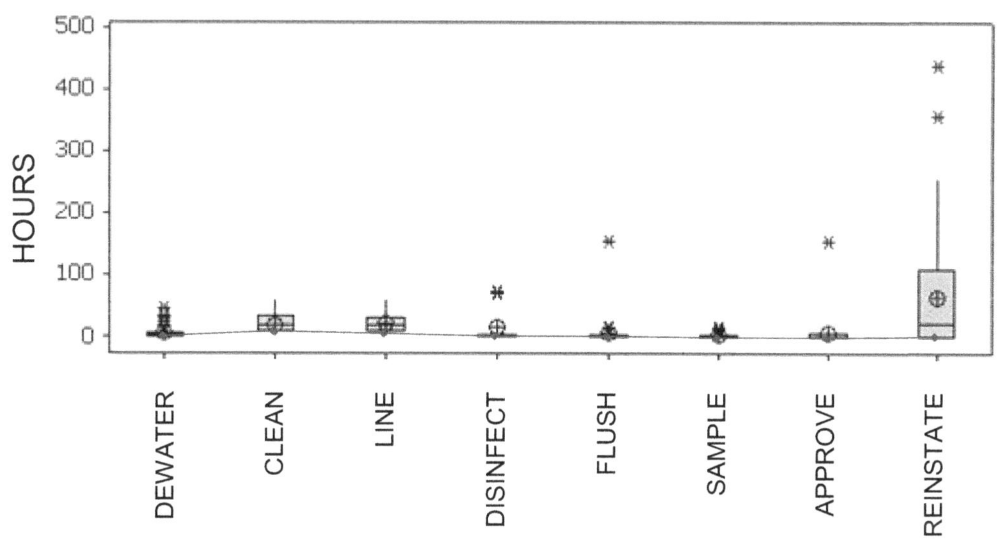

Figure 4.6: Comparison of Northern Kentucky Water District's cleaning and lining operation by man hour

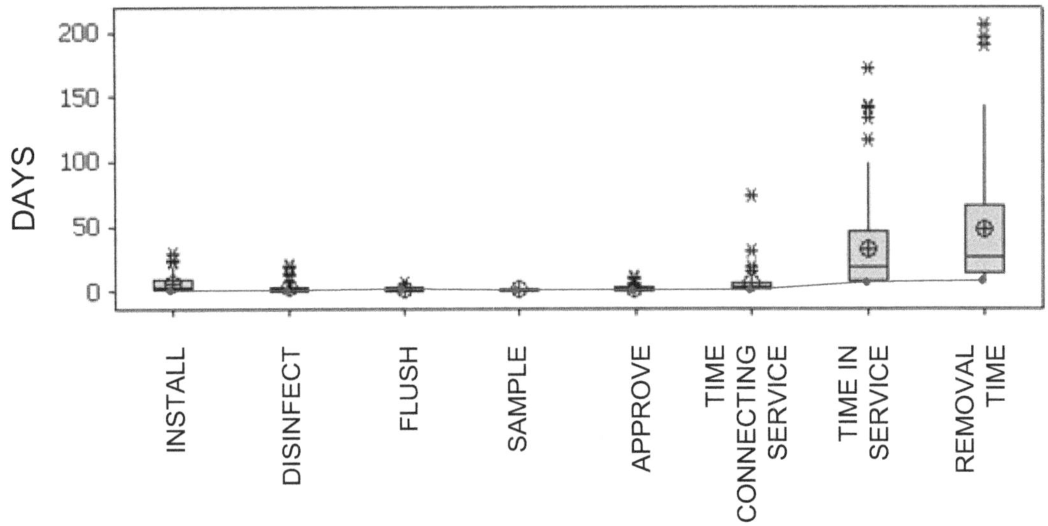

Figure 4.7: Comparison of Northern Kentucky Water District's temporary water supply operation by work day

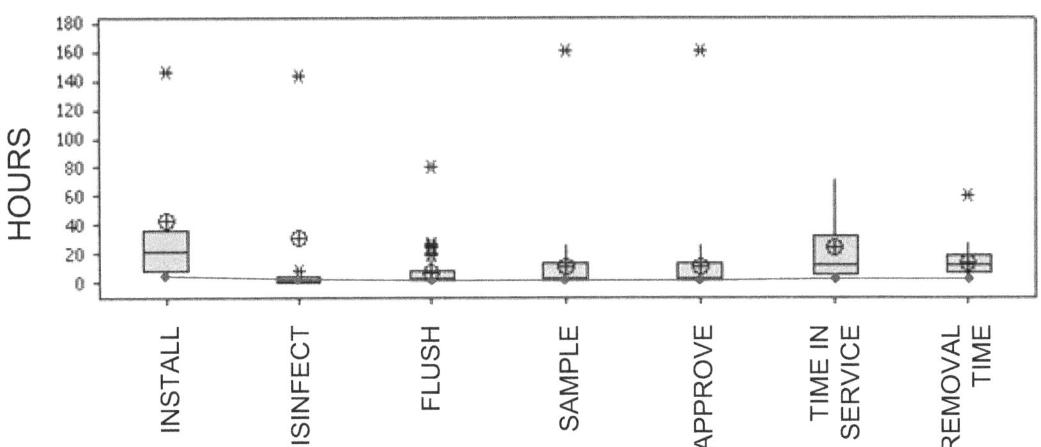

Figure 4.8: Comparison of Northern Kentucky Water District's temporary water supply operation by man hour

LOUISVILLE WATER COMPANY

Program Background

The Louisville Water Company (LWC), located in Louisville, Kentucky provides water to approximately 238,200-residential service taps and 22,650–commercial and industrial accounts in the Louisville metropolitan area and supplies retail and wholesale water to 4-adjacent Counties through 11-separate master meter locations. The average 125-million gallons per day consumer demand is met by two drinking water treatment plants that use the Ohio River as their source.

Water service is provided to the residential, industrial and wholesale customers via a 3,700 mile water transmission and distribution pipeline system. This system includes pipe sizes ranging from ¾-inch service material to 60-inch transmission mains. The oldest know pipe still in service is 20-inch diameter, unlined cast iron pipe installed in 1862. It is estimated that 8% of the current pipeline network, about 300-miles, was installed as unlined cast iron pipe between the mid-1800s until about 1932. Since 1932, the LWC has primarily installed lined cast iron, PVC, PCCP, or HDPE materials for new pipe installation.

To help reduce low flows conditions, discolored water complaints, and improve inadequate fire protection, the LWC initiated a water main cleaning and in situ lining program in 1985. Cement mortar lining was selected as the optimum rehabilitation technique for the LWC system. The rehabilitation program began with the larger transmission mains and ultimately will address all structurally sound, unlined cast iron. As such, the cleaning and lining program has adapted its application to accommodate pipelines ranging from 48-inch to 4-inch diameter.

Unlike many other utilities, LWC's capital budget for water main renewal was a fixed amount for each year of the multi year program. Each year's lining work was generally bid as one large project to encourage economies of scale and to increase contractor competition. This consistency of work further reinforced the cost effectiveness of the program by guaranteeing a known quantity of work on Louisville's system annually. During the height of the program LWC lined 25-miles of main each year.

Customer Relations

The LWC implements a multi-level contact approach to notify customers of pipe lining work and to inform them of planned water outages in their area. Approximately 2-weeks prior to any construction or contractor mobilization, the LWC Project Manager sends affected customers letters and information notifying them of the upcoming project. Included in the information packet is a brief letter with an explanation of the construction process, the Project Manager's contact information, emergency contact information, a map of the project area, a brochure that explains the importance of maintaining waterline infrastructure and a statement of LWC's commitment to providing the best possible water quality and service to its customers.

As the pipe lining work moves closer to an individual customer, LWC policy dictates that the contractor will make every effort to notify each customer prior interrupting service during transfer to the temporary supply and again when reinstating the permanent service line. Customers are also given 24-hour notification prior to planned outages that require water to be shut off for extended periods of time.

Typical Project Construction

A typical annual cleaning and lining project for LWC would consist of an average of 25-miles of pipeline ranging in size from 4-inch diameter to 24-inch diameter. Often, the larger pipes did not have service connections and therefore did not require installation of temporary pipe. This allowed for them to be taken out of service and lined during winter months. When weather permitted temporary pipe was installed so that the smaller distribution mains could be addressed.

A year's lining effort was generally divided among only four to six neighborhoods within Louisville's aging system. Each area contained enough footage to permit the contractor to remain mobilized at a site and to develop standard process routines. By distributing the work in several large sections, the contractor was able to provide a cost effective price and work quickly through the neighborhoods.

A temporary water supply is provided to every customer whose existing service will be affected by the construction. The majority of the temporary pipe installed is 2-inch diameter PVC. Occasionally, for large domestic services, it is necessary to install 4-inch and sometime 6-inch diameter pipeline to meet flow demand. Close coordination with local fire districts and rapid return to service processes permit fire hydrants to be taken out of service without providing alternative or temporary hydrants. The temporary pipe is typically buried flush with roadways at crossing and ramped with rubber ramps at customers' driveways. Louisville also requires the pipe to be buried at commercial entrances where it will be exposed to heavy truck traffic.

After the temporary pipe is installed, it is filled and pressurized to inspect for leaks. After a successful leak test the pipe is drained and refilled with 50ppm hyperchlorinated water. Louisville requires that the hyperchlorinated water remain in contact with the pipe for a minimum of 24-hours before being flushed. Following the 24-hour contact time (CT) the water is flushed and a sample taken to the Company's laboratory to perform bacteriological tests. Sample incubation requires another 24-hour time period. This entire chlorination and laboratory testing process is repeated every time the temporary pipe is depressurized and reinstalled at a new location.

Subsequent to laboratory approval connections to individual services are installed. Flexible hose or tubing is used to connect the temporary pipe to customers' meters. Only that tubing and hose which has been approved by the Food and Drug Administration (FDA) and National Sanitation Federation (NSF) for contact with potable water is permitted. Upon connection of all customer service lines to the temporary pipeline the cleaning and lining process may commence.

Study Area Description

Data for this report analysis originated from a combined commercial and residential area. The distribution system within this area consisted of an unlined cast iron gridded network system with relatively few dead ends. Data was obtained from eight separate cleaning and lining runs for a total 3,000 linear feet of lining. Pipe diameters of the rehabilitated mains ranged from 6-inches to 20-inches. During the rehabilitation, water service was maintained to 50-service connections by placing 6,300 linear feet of 2-inch diameter through 6-inch diameter PVC temporary pipe.

Task Rate Comparison

Figures 4.9 through 4.12 display Louisville Water Companies pipe rehabilitation program as compared to other utilities participating in the study. The data are segregated by lining and temporary service tasks and also by man hours and days per task. The solid line running horizontally across the figure represents LWC data. The background box plot represents the average utility data as analyzed and presented in Chapter 3.

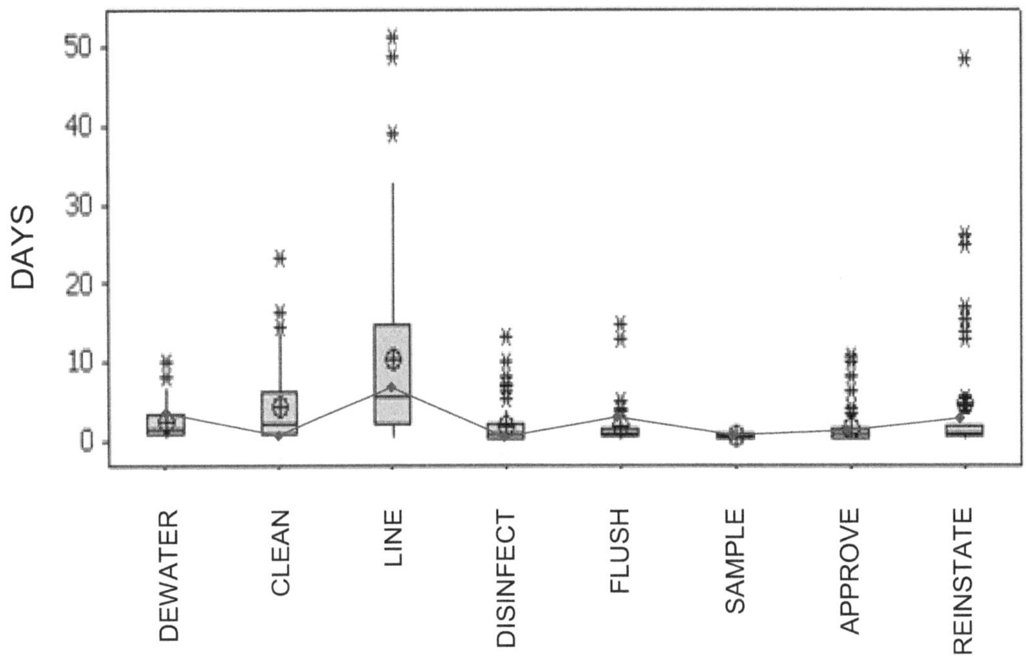

Figure 4.9: Comparison of Louisville Water Companies cleaning and lining operation by work day

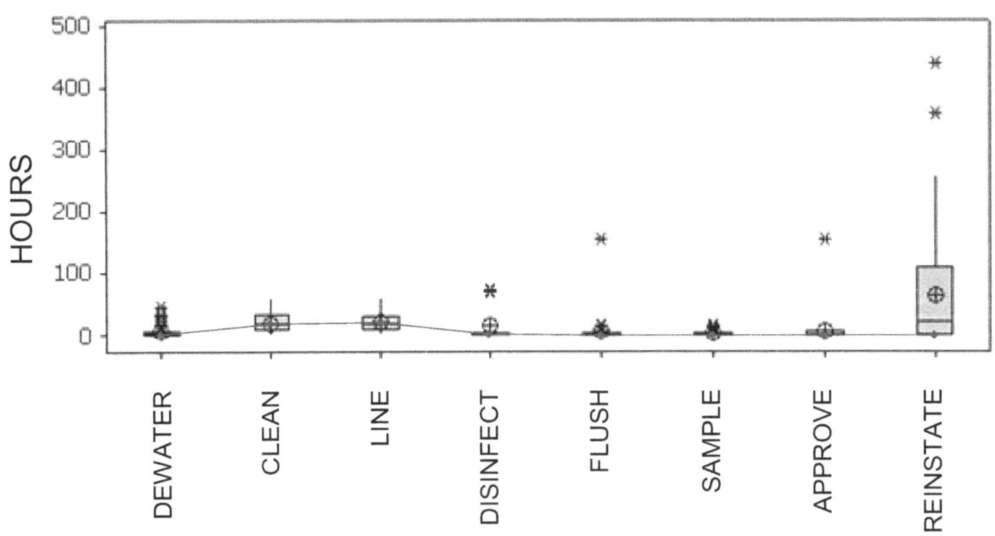

Figure 4.10: Comparison of Louisville Water Companies cleaning and lining operation by man hour

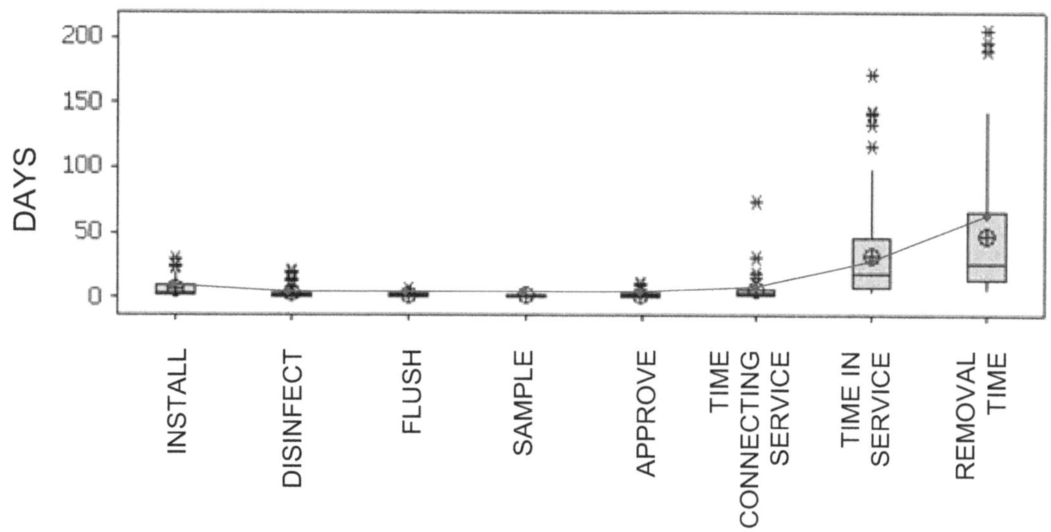

Figure 4.11: Comparison of Louisville Water Companies temporary water supply operation by work day

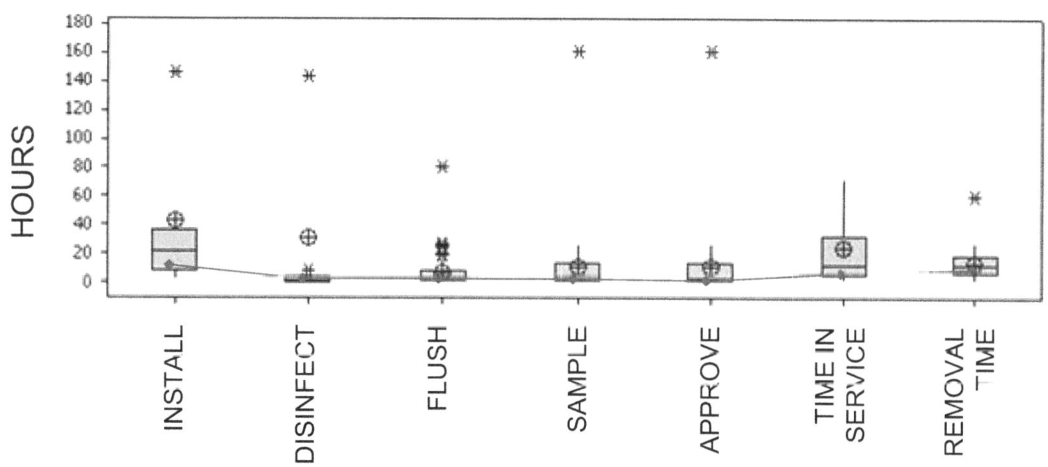

Figure 4.12: Comparison of Louisville Water Companies temporary water supply operation by man hour

OMAHA, NEBRASKA – MUNICIPAL UTILITIES DISTRICT (CMU)

Program Background

Omaha, Nebraska Municipal Utilities District (CMU) serves approximately 96-million gallons of water each day to 180,500 customers through a 2,507-mile transmission and distribution pipeline network. The first pipe known to have been installed in Omaha was in 1881. Initially, Omaha primarily installed unlined cast iron pipe, however, after about 1949 Omaha began purchasing pipe that was cement mortar lined at the foundry. Within their distribution system, it is estimated that approximately 257-miles, are unlined cast iron and 1,028 are line cast iron. The pipes in the Utility District's system range in size from 4-inch diameter to 48-inch diameter.

Since 1960 CMU has been working to line the unlined cast iron pipe within it's distribution system. Driving factors for initiating the lining program were to eliminate red or discolored water complaints and to improve fire flows. The initial lining program continued for 15-20 years and resumed again in the 1990s. Currently the cleaning and lining program is structured on a two year cycle where design work is completed in year one and project construction in year two. The cycle begins again with new design work in year three.

The data for this report was collected during the most recent cleaning and cement mortar lining project in Omaha. This project was carried out in a gridded City setting that contained a mix of residential and commercial customers. The project represented approximately 2% of the Omaha's unlined pipe network. The project scope included 4- through 16-inch diameter pipe.

Task Rate Comparison

Figures 4.13 through 4.16 display Omaha's pipe rehabilitation program as compared to other utilities participating in the study. The data are segregated by lining and temporary service tasks and also by man hours and days per task. The solid line running horizontally across the figure, represents Omaha's data. The background box plot represents the average utility data as analyzed and presented in Chapter 3

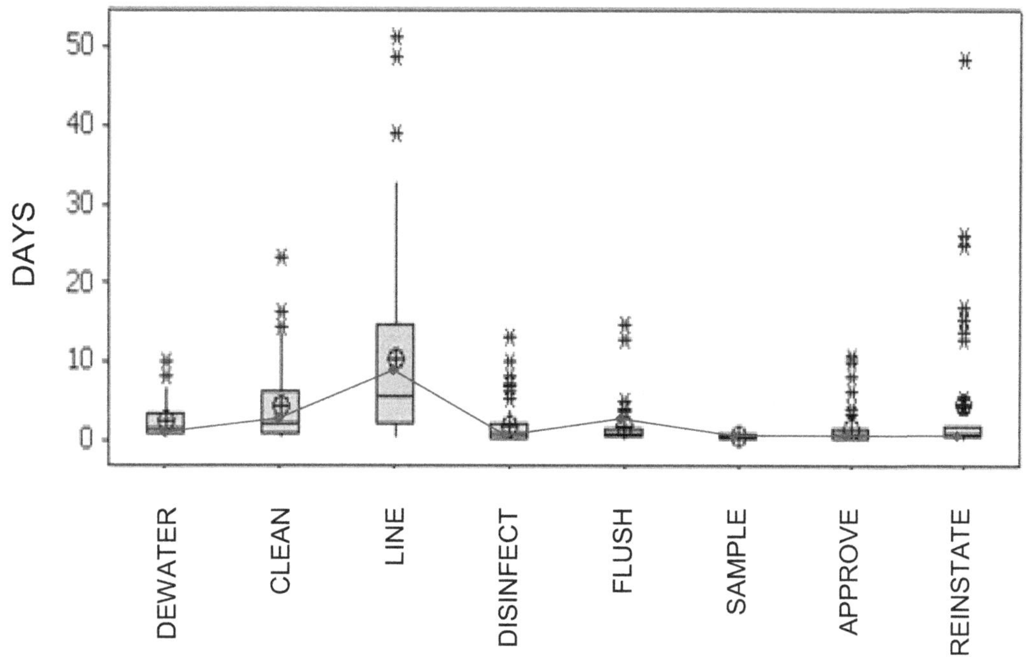

Figure 4.13: Comparison of Omaha's cleaning and lining operation by work day

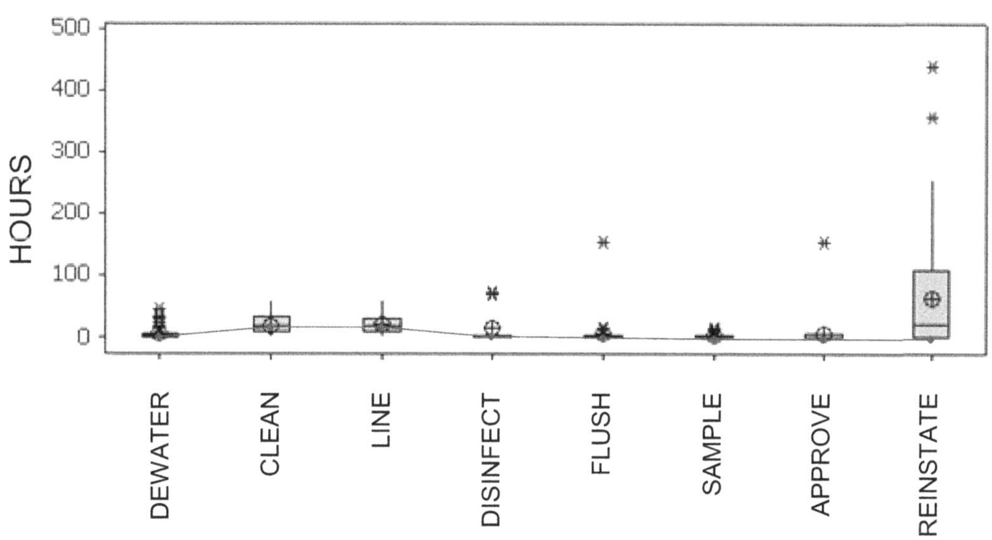

Figure 4.14: Comparison of Omaha's cleaning and lining operation by man hour

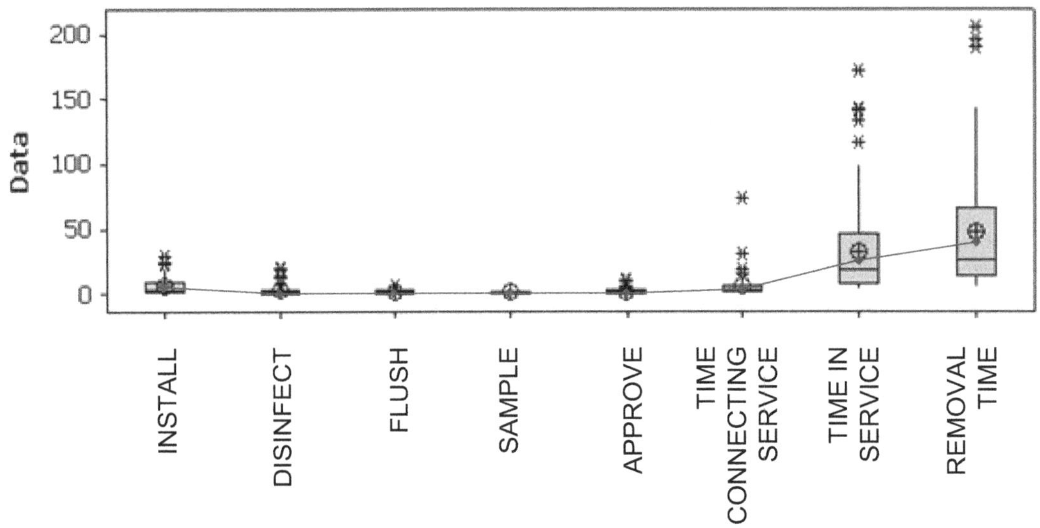

Figure 4.15: Comparison of Omaha's temporary water supply operation by work day

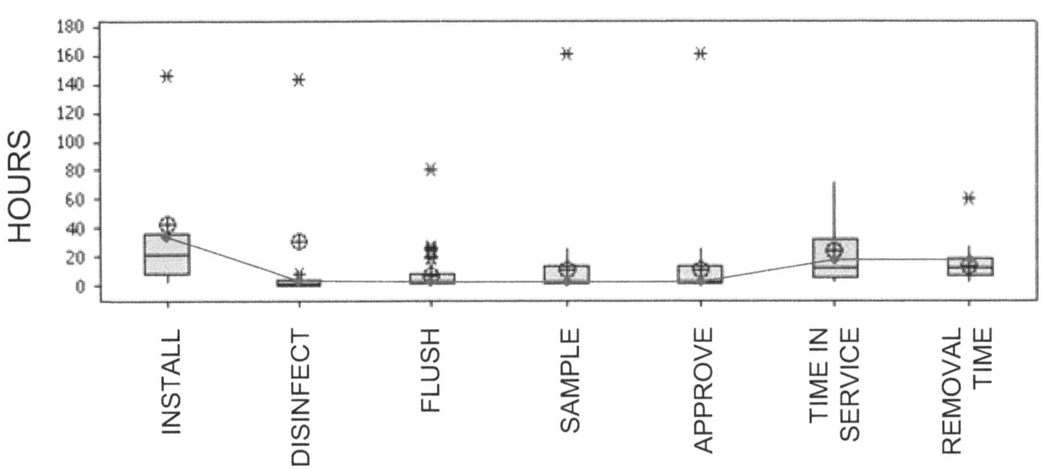

Figure 4.16: Comparison of Omaha's temporary water supply operation by man hour

CLEVELAND DIVISION OF WATER

Program Background

The Cleveland Division of Water, supplies potable water the City of Cleveland and its neighboring suburbs. The system serves approximately 1.5-million people an average of 400-million gallons per day and uses Lake Erie as its source.

Cleveland's 5,620-mile water distribution system is composed of 420-miles of large diameter transmission main and 5,200-miles of smaller diameter distribution main. Like many cities, Cleveland began installing unlined cast iron pipe in the 1880's. The utility continued to install unlined materials until the 1950's when they began purchasing factory lined cast iron pipe. To date, Cleveland has not been able to accurately determine the amount of unlined cast iron water main within it's system.

Water main cleaning and in situ lining has been a routine part of Cleveland's capital improvement program since the 1950s. Primary factors for implementing a cleaning and lining program were to eliminate low flows, reduce discolored water complaints, and improve inadequate fire protection. Each year Cleveland lines approximately 20-miles of main through two large projects. It is believed that the project size encourages economies of scale and facilitates contractor competition by guaranteeing a known quantity of work on Cleveland's system annually.

Customer Relation

Cleveland designates between six and eight project areas for cleaning and lining each year. As areas are identified, Cleveland Division of Water notifies the elected council members in each of the affected areas. The council members are responsible for conveying the information to the residents at public meetings. A conscious effort is made to distribute the renewal work among all the wards throughout the City.

An additional notice is sent to each affected customer as construction moves into new areas. The notice explains the need for the temporary pipe, what to expect in the way of inconveniences, and the project duration. Cleveland's policy is to drop the meter from service during the time its customers are on temporary supply and an explanation of the billing policy and the fact that the customer will not be billed for water used during construction is also included.

Cleveland requires its Contractor to have a 24 hour contact number for any emergency including temporary pipeline breaks. A complaint log is maintained by the contractor on electronic format and submitted to City on a weekly basis. Entries on the complaint log include: problems and complaints, reported date, response date, action taken, etc.

Typical Project Construction

The initial step in the lining process is to install temporary a pipeline network so as to maintain almost continuous service to customers. For each street affected, a 4-inch temporary main will be installed on the side with existing fire hydrants and a 2-inch temporary pipe installed on the opposite side. A temporary fire hydrant connection is provided for each permanent fire hydrant removed from service during construction. The temporary lines are thoroughly flushed and sampled for water quality; they are not, however, hyper-chlorinated prior to sampling. Two consecutive negative samples, 24-hours apart, are required before customers can be connected to the temporary main.

Once laboratory testing confirms water quality, hoses are run from the temporary pipe to individual customers. In the absence of outside meter settings the customers' connection is made to an outside spigot or sill cock. Generally, Cleveland's meters are set inside customers' homes or businesses and a substantial coordination effort is required to secure the temporary supply. Generally, the responsibility for making arrangements to meet with customers is entirely the contractors.

After the temporary network is established, the cleaning and lining process is initiated. Currently, Cleveland use of a drag scraper to clean the lines. Once cleaned, cement mortar is spray applied to the prepared lines.

Cleveland gives contractors the option to use either air or water to blow back services following lining. If water is used the contractor will operate either the outside curb stop or the inside main line shut off valve. If air is to be used the contractor must access the service line at either outside at the meter vault or inside at the main line shut off or the meter setting. For Cleveland's system air blowbacks seem to be less destructive to the new lining than water blowbacks, especially on small diameter mains.

Study Area Description

Data for this report analysis originated from a mixed-use commercial and residential area. The distribution system within this area consisted of an unlined cast iron gridded network system with relatively few dead ends. Data was obtained from ten separate cleaning and lining runs for a total 16,900 linear feet of lining. All ten segments were 6-inch diameter cast iron. During the rehabilitation, water service was maintained to 585-service connections by placing 37,150 linear feet of 2-inch and 4-inch diameter HDPE temporary pipe.

Task Rate Comparison

Figures 4.17 through 4.20 display Cleveland's pipe rehabilitation program as compared to other utilities participating in the study. The data are segregated by lining and temporary service tasks and also by man hours and days per task. The solid line running horizontally across the figure represents Cleveland's data. The background box plot represents the average utility data as analyzed and presented in Chapter 3

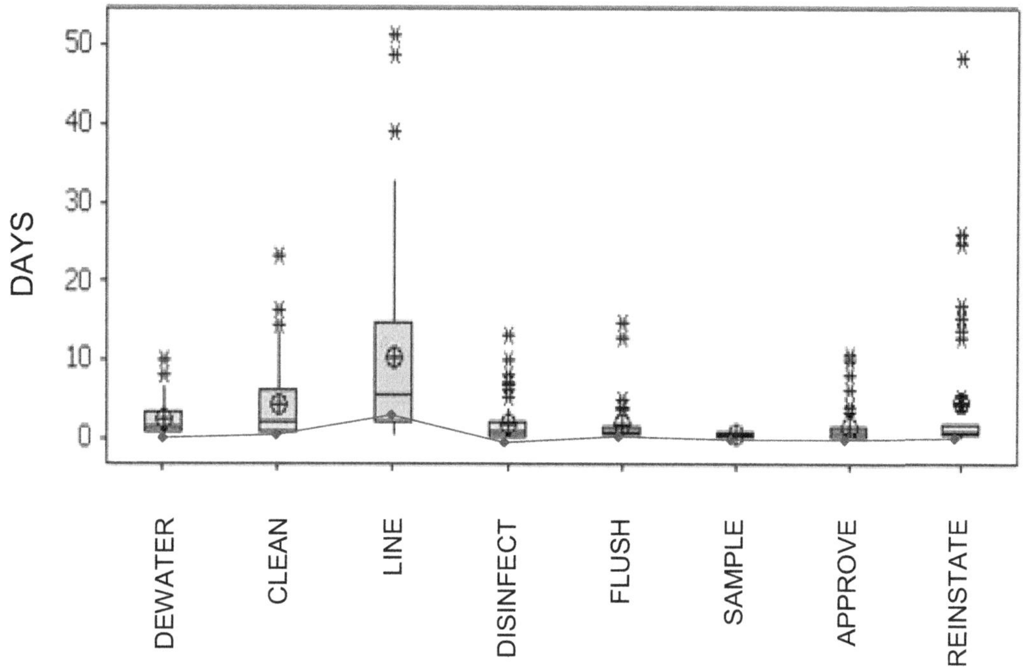

Figure 4.17: Comparison of Cleveland's cleaning and lining operation by work day

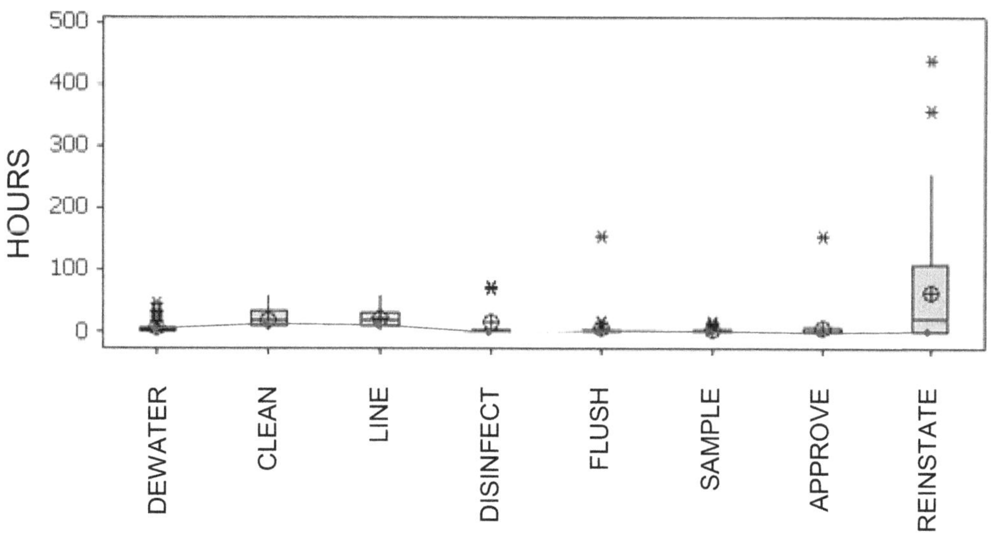

Figure 4.18: Comparison of Cleveland's cleaning and lining operation by man hour

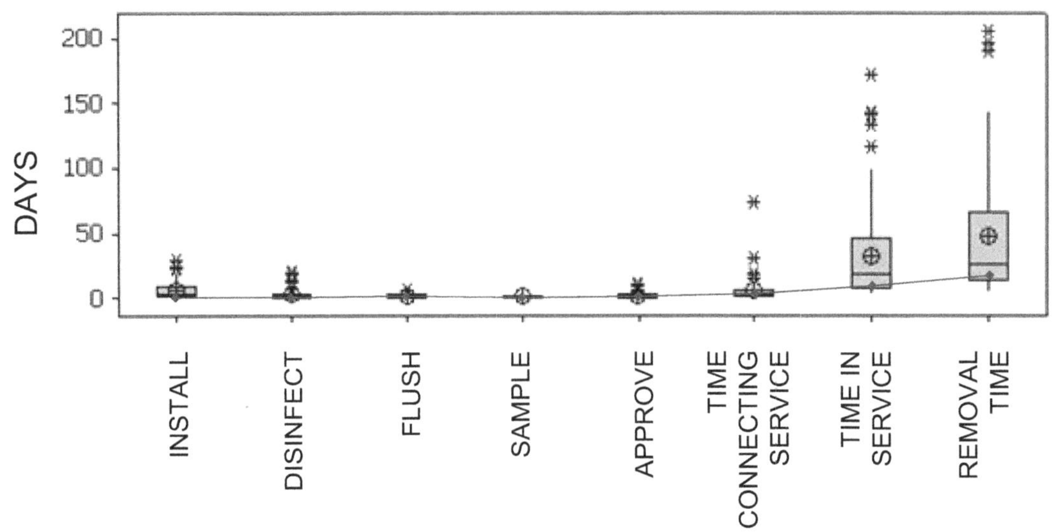

Figure 4.19: Comparison of Cleveland's temporary water supply operation by work day

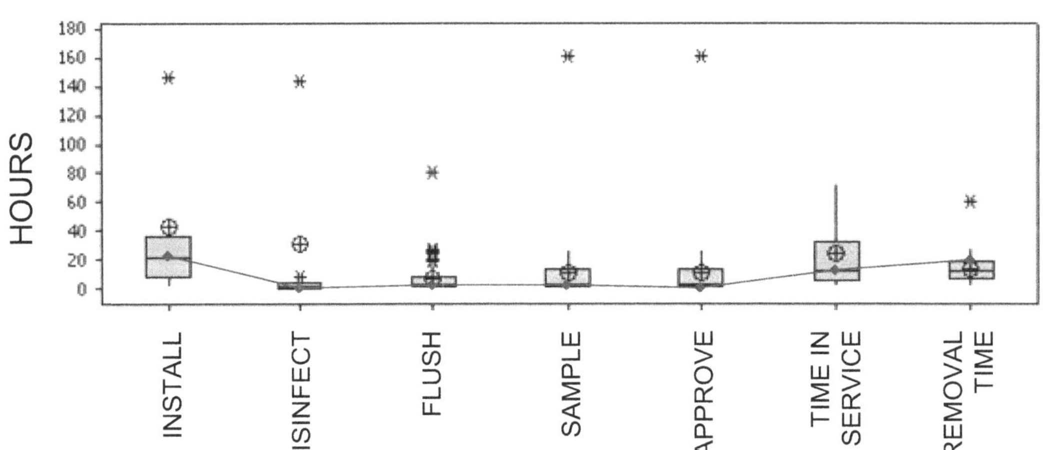

Figure 4.20: Comparison of Cleveland's temporary water supply operation by man hour

GREATER CINCINNATI WATER WORKS

Program Background

The Greater Cincinnati Water Works (GCWW), located in Cincinnati, Ohio provides water to approximately 224,000 accounts of residential service taps and 10,500 industrial taps. GCWW also supplies wholesale water to Butler and Warren Counties , Ohio and to Boone County, Kentucky. The average 136-million gallon daily demand is provide by the Miller Plant the Bolton Plant. In addition, GCWW operates the Mason Treatment Plant, which was placed in stand-by service in November, 2004.

Water service is provided to the 900,000 residential, industrial and wholesale customers via a 3,050-mile water transmission and distribution pipeline system. The first pipe installed in the Cincinnati water system was in the early 1800's; however, the oldest pipe still in service is a 10-inch diameter, unlined cast iron pipe installed in 1850. It is estimated that 30% of the current pipeline network (approximately 1,000 miles), was installed as unlined cast iron pipe between the mid-1800s and 1947. Since 1947, GCWW has installed cement-lined cast iron and ductile iron, steel, PCCP, or HDPE materials for new pipe installation.

To improve hydraulic flow conditions in its unlined cast iron mains, GCWW initiated a water main cleaning and cement mortar lining program in the 1960s. The program continued for several yeas with marginal performance and was finally discontinued in the mid 1970s. In 1999 GCWW completed a pilot cement mortar lining program in structurally sound, transmission mains with good results. Subsequently, the cleaning and cement mortar lining program has been reinitiated to take advantage of more competition and newer technologies. To date, the cleaning and lining process has been used in three projects with transmission pipelines ranging from 24-inch to 60-inch diameter and two distribution projects of 6-inch, 8-inch, and 12-inch pipelines with one project still ongoing.

GCWW's rehabilitation program is in the developmental stage and typical or routine project construction administration has not been adopted. Currently GCWW is evaluating various aspects of pipeline rehabilitation and replacement processes to make a determination on how best to implement pipeline renewal.

Customer Relations

GCWW's first efforts to rehabilitate distribution mains ended with mixed results. Although the technology and final product were acceptable, the public perception of the project was generally poor. Several issues contributed to the poor results including an overall lack of communication between GCWW and its customers and vague project specifications. Primary among the customer concerns was that the temporary piping was left in place for extended periods of time, improperly installed, and the construction activities took longer than originally anticipated.

Prior to the second distribution project being bid GCWW hired a communications consultant to assist them with customer related issues. Several key changes were made and implemented as a result of the specialist's involvement. GCWW published informational brochures and FAQ sheets, which have been used extensively and include all aspects of the project from water quality to what the streets would look like during construction. Community meetings were initiated long before letting of the contract to inform the residents as well as local fire and police. Once the project commenced letters were hand delivered to the affected homes and businesses, which included the brochure and FAQ sheet. This gave GCWW an opportunity to provide face-to-face contact and answer questions. Specialized door hangers were also developed and have been used throughout the project area for planned outages.

More detailed specifications related to customer service issues and increased involvement of the construction inspection office in resolving customer complaints in a timely manner were put in place. The improved project specifications detailed GCWW's policies and procedures for interaction between contractor personnel and customers, resulting in clearer contractor expectations. Empowerment of the construction inspector to make field changes resulted in fewer initial customer complaints as well as more timely complaint resolution. In addition to full time inspection GCWW also employed the services of a consultant who has over 35 years of experience in the water industry to complement and facilitate the work.

Typical Project Construction

The GCWW cleaning and lining program actively accommodates customer reactions into its work scheduling. Rehabilitation projects involving distribution mains, and subsequently a large number of customers, are scheduled for warmer times of the year when temperatures rarely drop below freezing and thus the temporary water supply pipe is protected. Alternatively, lining of large diameter or transmission mains is commonly performed during the winter months when installation of at grade temporary pipe is not practicable.

When a temporary water service is required, the temporary pipes are installed at grade, chlorinated, flushed, and the water sampled for bacteriological testing prior to making any customer connections. Every effort is made to notify each customer immediately prior to being connected to the temporary piping. The most desirable connection point to the customer's service is through the hose bib or sill cock. In the event that this connection point is not available or inoperable the contractor is required to excavate the service line at the curb stop and make the temporary connection at that point. The least desirable situation is to make the connection inside the premise. GCWW's policy is to exhaust all alternative connection points prior to entering the premise to make the connection. During the pipeline renewal project all problematic shut off valves are replaced.

Once the temporary pipe network is installed and the chlorination procedures are completed, the main to be rehabilitated is isolated from the system and the cleaning process is initiated. GCWW prefers the drag scrap cleaning method, but has allowed a small amount of hydraulic cleaning in its transmission main system. After the main is cleaned it is inspected via video camera and the contractor submits the tape to GCWW for review. When possible on site personnel witness the video inspection work to speed the approval process. Lining operations follow as closely behind the cleaning process as is practical. As GCWW does not isolate the homes due to internal meter settings, clearing cement mortar from each service line, referred to as blowbacks, are problematic. Service blowbacks are not always effective due to the need to reverse the flow through the meter. GCWW maintains customer service representatives on the project site to facilitate the contractor's work, to make internal inspections, and in some cases replace meters. After the main is lined another video inspection is made and tape submitted. After viewing the video and approval of the lining product, all improvements are made to the main, including valves, fire hydrants, and connections between segments.

Chlorination Procedures

All temporary piping and newly lined mains are disinfected with chlorine, flushed, sampled and tested for bacteriological contamination prior to being put into service. The pipelines, whether temporary or permanent, are hyperchlorinated at 25 ppm concentration and maintained for a 24-hour contact time. Following the 24-hour contact period the main is flushed to achieve system level chlorine concentration (between 1 and 2 ppm). Water samples are obtained and incubated for 24-hours at GCWW's laboratory. A second set of samples is taken 24-hours after the first and also tested at the laboratory. If either sample fails, the test is repeated.

During the chlorination process, GCWW is concerned that there is no positive separation between the inside meters and the hyperchlorinated pipe. As such, a depressurization of the temporary pipe could create a vacuum and draw hyperchlorinated water from the newly lined main into the house service lines. To quickly react to this potential problem, GCWW requires the contractor to maintain at least 1-employee on-site throughout the hyperchlorination contact time. In addition, GCWW requires the contractor to provide 24-hour on-site police presence to prevent tampering and to provide additional observation of temporary pipe in the event a depressurization occurs during the chlorination phase.

Study Area Description

Data for this report analysis originated from GCWW's second distribution system project which includes a combined commercial and residential area. The distribution system within this area consisted of an unlined cast iron gridded network system with relatively few dead ends. Data was obtained from fourteen separate cleaning and lining runs for a total 6,200 linear feet of lining. Pipe diameters of the rehabilitated mains ranged from 6-inches to 12-inches. During the rehabilitation, water service was maintained to 50-service connections by placing 6,300 linear feet of 2-inch diameter through 6-inch diameter PVC temporary pipe.

Task Rate Comparison

Figures 4.21 through 4.24 display GCWW's pipe rehabilitation program as compared to other utilities participating in the study. The data are segregated by lining and temporary service tasks and also by man hours and days per task. The solid line running horizontally across the figure, represents GCWW's data. The background box plot represents the average utility data as analyzed and presented in Chapter 3.

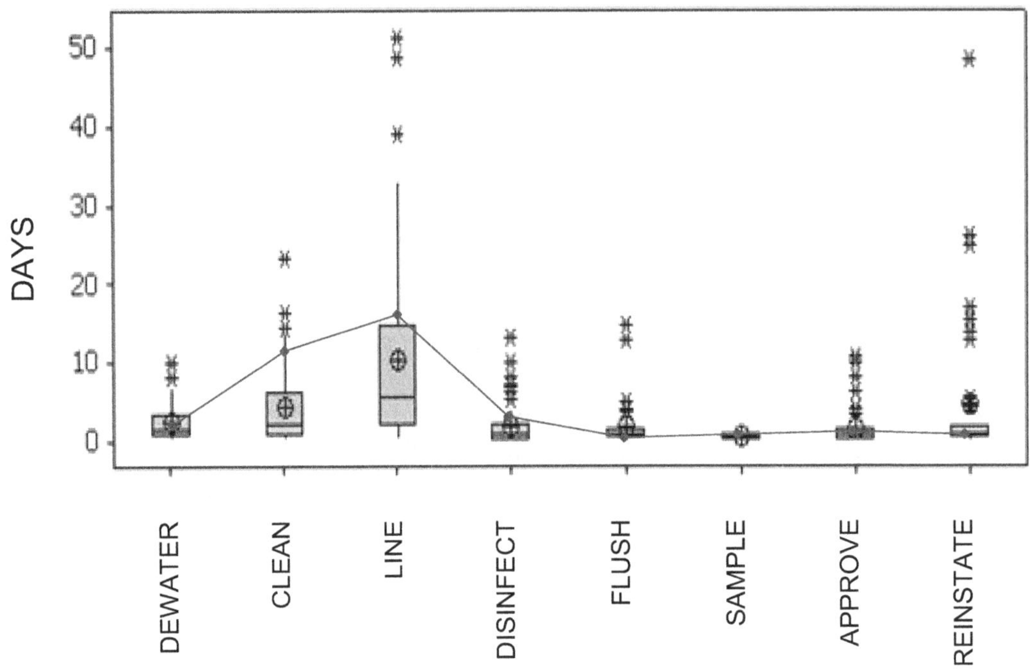

Figure 4.21: Comparison of GCWW's cleaning and lining operation by work day

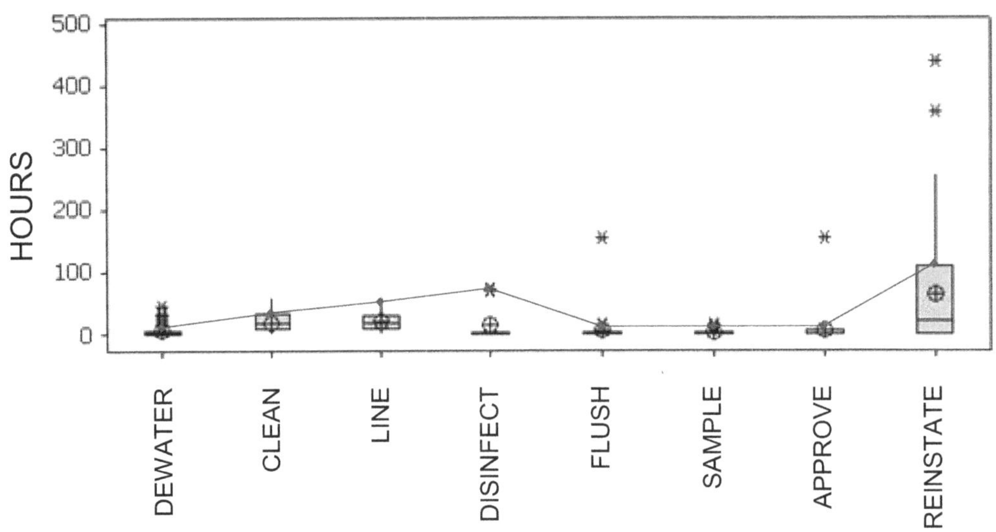

Figure 4.22: Comparison of GCWW's cleaning and lining operation by man hour

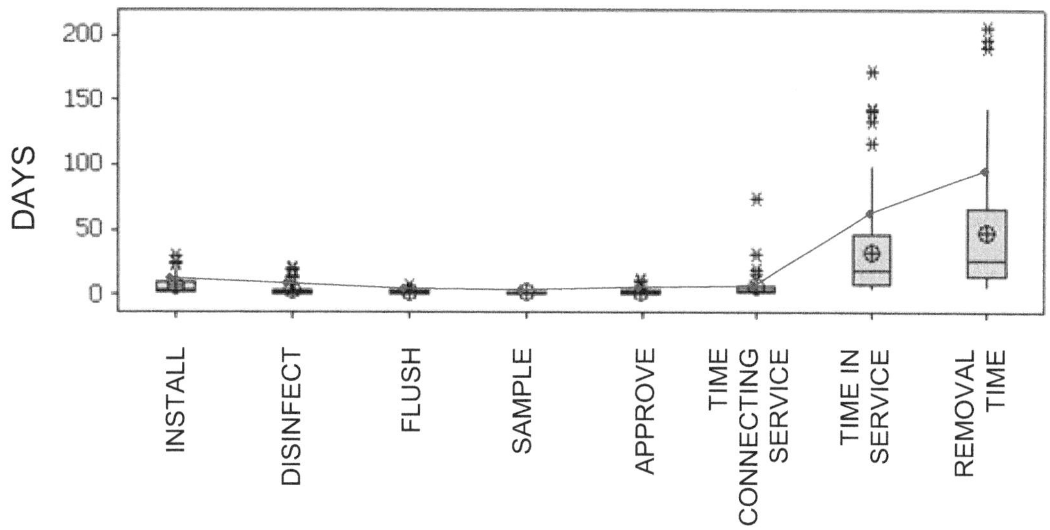

Figure 4.23: Comparison of GCWW's temporary water supply operation by work day

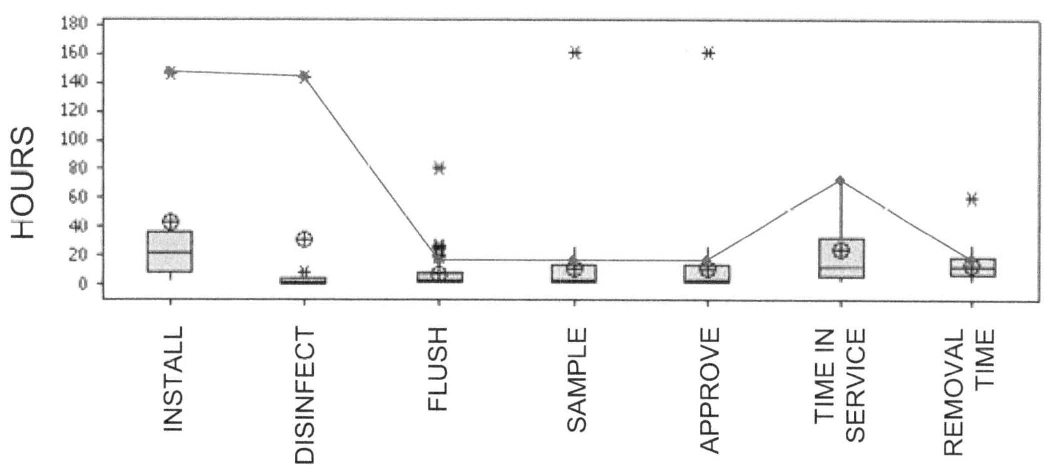

Figure 4.24: Comparison of GCWW's temporary water supply operation by man hour

CHARLOTTE-MECKLENBURG UTILITIES

Program Background

Charlotte-Mecklenburg Utilities (CMU), located in Charlotte, North Carolina provides water to approximately 220,000 service taps in the metropolitan area and supplies retail and wholesale water to 2 adjacent Counties. The average 105-million gallons per day consumer demand is met by three drinking water treatment plants that use Mountain Island Lake and Lake Norman as their source. Water service is provided through a 3,400 mile transmission and distribution system ranging in size from 3/4-inch services to 60-inch transmission mains. There are approximately 1,000 miles of unlined cast iron pipe in the system.

CMU's cleaning and lining program was implemented in 1998 to address water quality and fire flow issues in the older parts of the distribution system. The focus of the lining effort has been on 6- and 8- inch mains as marked improvement can be obtained. A pilot project using cement mortar lining was conducted in 1998, but after evaluation of available technologies epoxy lining was selected due to water chemistry concerns.

CMU's current water rehabilitation budget is $7 million annually and approximately 2.5 million of that is used for cleaning and lining. One- or multi-year unit price contracts are issued and the contractor is then directed to existing or newly developing problem areas.

Customer Relations

Prior to any construction activity, CMU implements a citizen notification policy with multiple contact methods. For lining activities, the notification procedure is modified by including a description of the lining process, the temporary water system setup, and providing emergency contact information. Customers are provided a 72-hour notice of any extended water interruption. Local fire officials are also notified and provided with maps of the proposed temporary water system network.

Typical Project Construction

The typical CMU epoxy lining project is performed through an annual unit price contract and consists of all work necessary to mechanically clean and epoxy line 6- and 8 –inch cast iron mains. Temporary service is provided to all customers during the work through an above ground network of PVC water lines. Prior to customer connection, at the meter box the temporary service network is disinfected following procedure for new pipe installation. Access pits are excavated at strategic locations and the line is mechanically cleaned. The line is CCTV inspected after cleaning and again after lining for quality assurance. Valves and fire hydrants are replaced during the project. The lined main is disinfected following standard CMU procedure and returned to service. The epoxy lining process follows Water Research Center (United Kingdom) guidelines with some slight modification.

Study Area Description

Data was obtained from five separate cleaning and lining runs for a total 5,625 linear feet of lining. All of the rehabilitated mains analyzed were 6-inch diameter. During the rehabilitation, water service was maintained to 120-service connections by placing 11,950 linear feet of 2-inch diameter PVC temporary pipe.

Task Rate Comparison

Figures 4.25 through 4.28 display Charlotte's pipe rehabilitation program as compared to other utilities participating in the study. The data are segregated by lining and temporary service tasks and also by man hours and days per task. The solid line running horizontally across the figure, represents Charlotte's data. The background box plot represents the average utility data as analyzed and presented in Chapter 3.

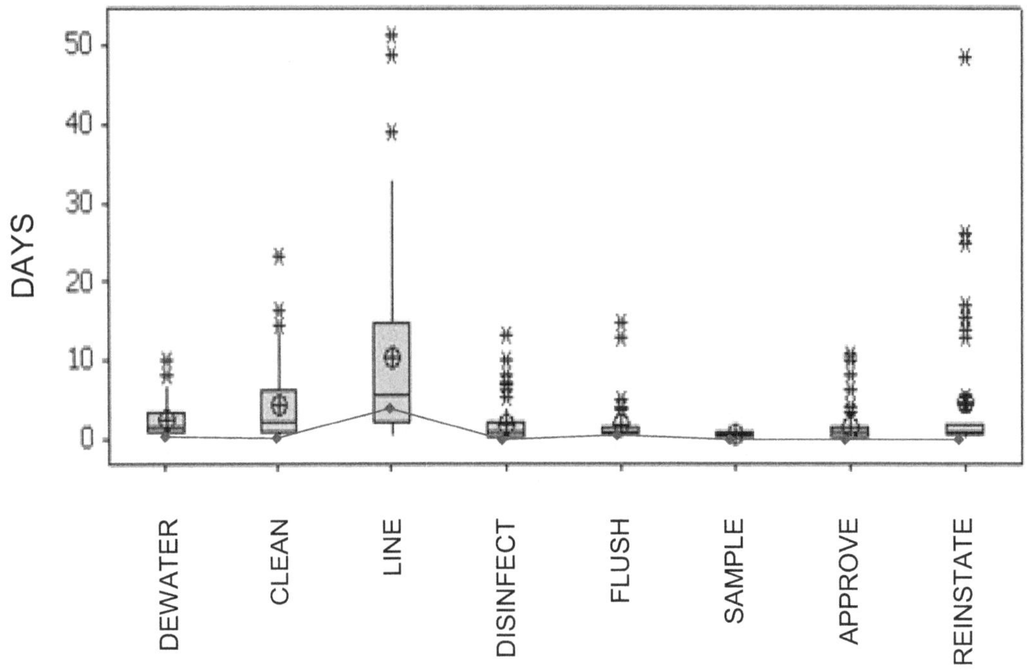

Figure 4.25: Comparison of Charlotte's cleaning and lining operation by work day

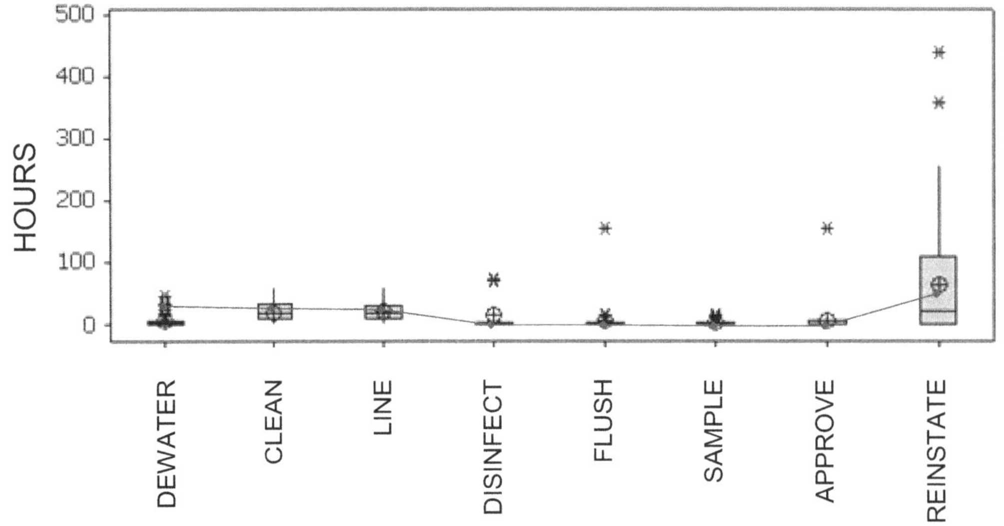

Figure 4.26: Comparison of Charlotte's cleaning and lining operation by man hour

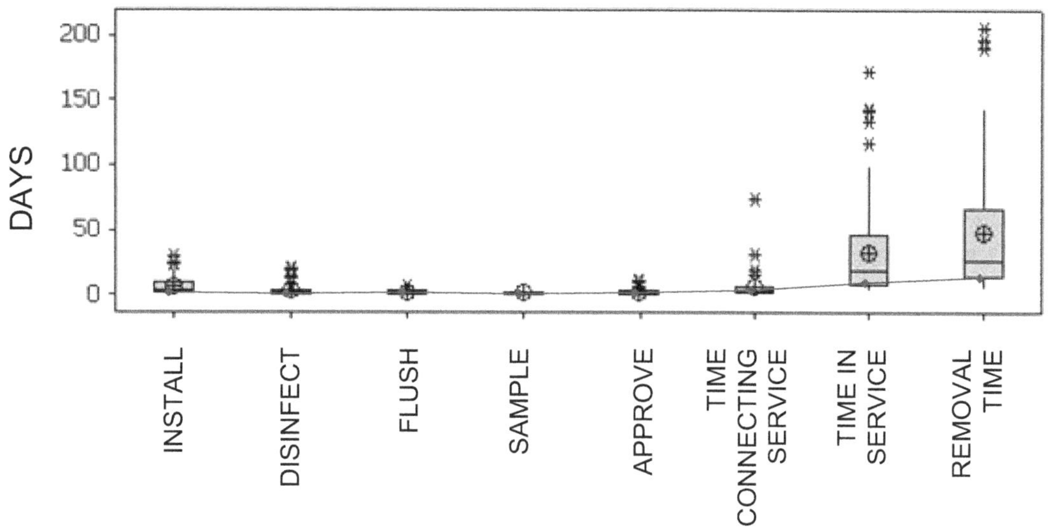

Figure 4.27: Comparison of Charlotte's temporary water supply operation by work day

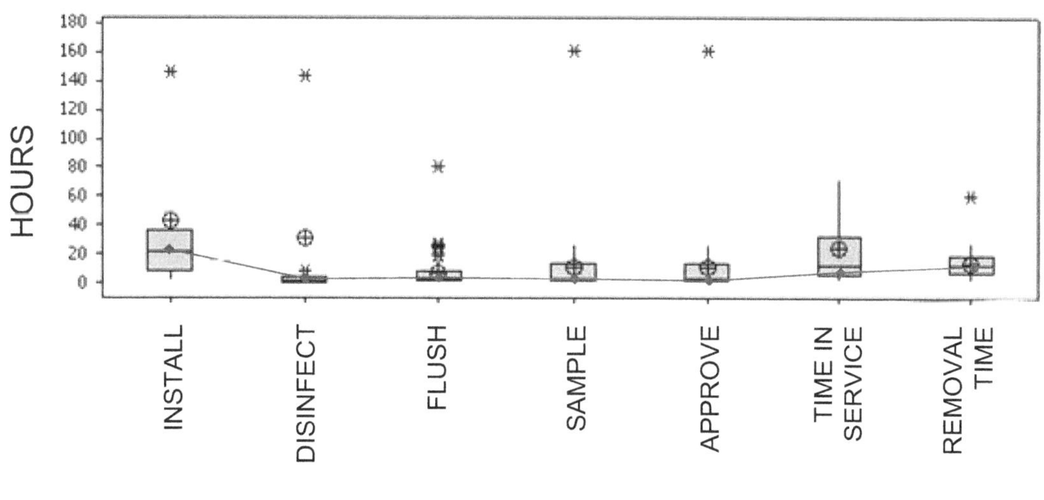

Figure 4.28: Comparison of Charlotte's temporary water supply operation by man hour

ANNE ARUNDEL COUNTY DEPARTMENT OF PUBLIC WORKS

Program Background

Anne Arundel County Department of Public Works (AACDPW), located in Anne Arundel County, Maryland provides water to a total of 92,000-customers. The drinking water supply system consists of 8-independent wells, 57-production wells, and 13 water treatment plants. Anne Arundel's system is capable of treating 37.5-million gallons of drinking water each day.

Water service is provided to the residential industrial customers via a 1,300 mile water transmission and distribution pipeline system. This system includes pipe sizes ranging from 2- to 16-inch mains. The oldest known pipe still in service is an unlined cast iron pipe installed in 1910. AACDPW continued to install unlined cast material until about 1960. It is estimated that there remains 100 miles of unlined cast iron pipe in the distribution system.

To help reduce low flows conditions, discolored water complaints, and improve inadequate fire protection, AACDDPW initiated a water main cleaning and in situ lining program in 1994. Cement mortar lining was selected as the optimum rehabilitation technique for the Anne Arundel County water system. The rehabilitation program has been ongoing for 10-years and ultimately will address all of the structurally sound, unlined cast iron pipe within the system.

Customer Relations

The AACDPW pipe rehabilitation program has generated formal procedures to educate and inform the public of infrastructure renewal operations. The AACPDW communicates its activities with the public in the following manner:

- The SER's (Systems Evaluation & Rehabilitation) projects are mapped and given to the county's Department of Public Works customer relations representative months ahead of the starts-of-work dates. The County's public relations representative is advised of the upcoming work and the project description is posted on the County's website.
- The public works customer representative contacts the community relations representative to discuss the nature and impact of the work on the residents and commercial customers. As needed, a public works representative attends community association meetings to address the public on the details of the project and possible disruptions/disturbances to service.
- At the preconstruction meeting the contractor designates a customer relations representative to act as the primary customer contact. The counties project inspector serves as a primary contact also. When possible both the contractor and inspector talk to the customer jointly.
- SER prints project specific door hangers as well as two important phone numbers. The county's emergency utilities dispatch number is listed principally for after-hours service calls. SER's office number is listed as the primary contact for information and complaints.

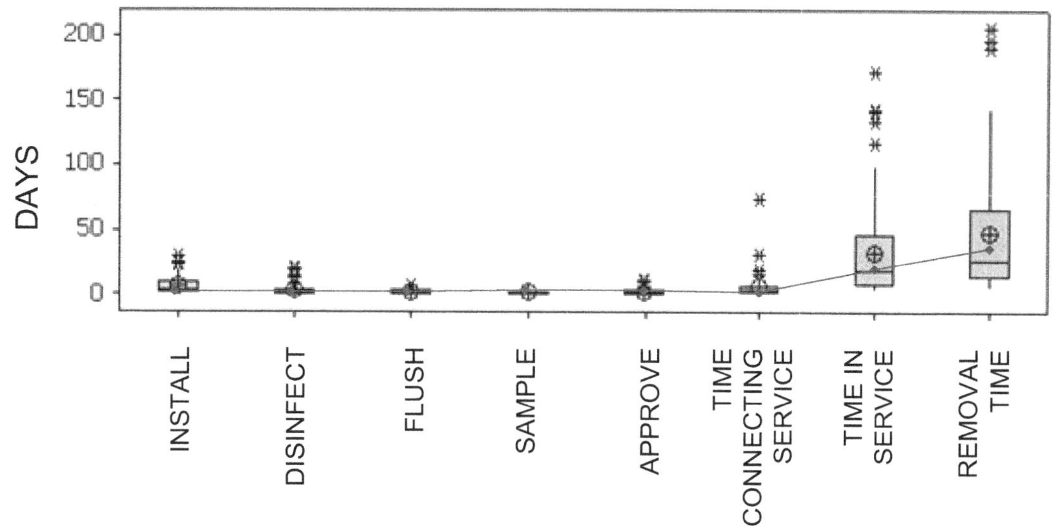

Figure 4.31: Comparison of Anne Arundel's temporary water supply operation by work day

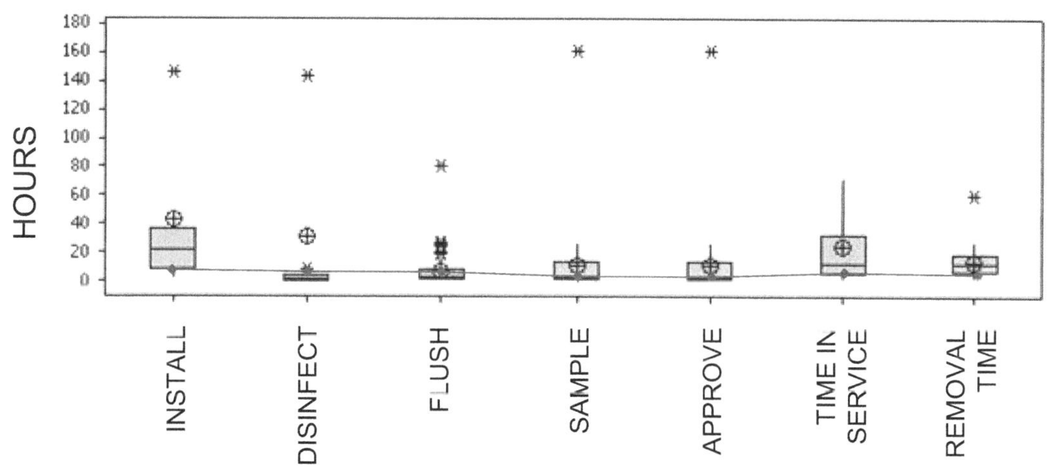

Figure 4.32: Comparison of Anne Arundel's temporary water supply operation by man hour

NASHVILLE PUBLIC WORKS

Program Background

Metro Water Services (MWS), located in Nashville, Tennessee provides water though approximately 156,000 service connections in the 388 square mile Nashville metropolitan area. The average 61.4-million gallons per day consumer demand is met by one drinking water treatment plant that relies on the Cumberland River as its source.

MWS's water transmission and distribution network is composed of approximately 2,700-miles of pipe, with diameters ranging from 2- to 60-inches. Nashville began installing factory cement mortar lined mains in the mid 1950s. Prior to that time the cast iron lines did not include protection against internal corrosion. About 33% of the system, approximately 900-miles, of unlined cast iron pipe was installed during this period.

MWS developed and implemented a Water Infrastructure Rehabilitation (WIR) program in 1998, to systematically address low flow rates and water quality issues resulting from unlined ductile iron pipe. MWS has concentrated on relining transmission mains that are 12 inches and larger as a first priority in the WIR program. To address mains on an equitable basis, a team was formed from various cross-divisional employees. The team created a project priority matrix to assess projects and set the rehabilitation order. The project priority matrix assesses pipeline rehabilitation needs based on:

- Area characteristics
- System characteristics
- Main characteristics
- Cost / benefit analysis

Each of these categories consists of a number of specific subcategories and an associated value. The higher the total value for a pipe segment the more likely it is to be considered for rehabilitation. Funding for the WIR program has varied every year but has averaged $ 6.2 million per year.

Customer Relations

Communicate with the public is an important part of Nashville's WIR program. A significant effort is made to not only keep the customer informed of the rehabilitation work, but to reduce the complications that the maintenance activities impart on local residence. To reduce customer complication's the WIR program aims to be sensitive to the needs of the community and the motoring public by:

- Adapting schedule(s) – Avoid work during traffic rush hours.
- Facilitating traffic flow – Keep one lane of traffic open.
- Using mobile and temporary signage
- Avoiding repeated maintenance operations within the same neighborhood

Prior to and during work activities, the WIR program strives to keep community members aware and involved in the rehabilitation effort. To meet this objective, WIR has established the following typical communication contacts:

- Briefings and courtesy calls for district Council Members, Mayor's Office of Neighborhoods, other departments and utilities. This allows for all the political entities to be well aware of a project before it happens.
- Notification letter to business and residential customers – This letter basically tells customers that MWS (and their respective contractors) will soon be in their area and why we are doing what is perceived to be by them a disruption to their daily routines.
- Community meeting letter – This letter invites not only the affected customers on the line that is to be rehabilitated but also everyone in the neighborhood
- Community meeting - At this meeting why, where, and how are also explained in detail.
- Door hangar - The Door hangar briefly describes the situation for the customers that happened to miss the community meeting.
- Shutdown letter – This is a letter that explains the shutdown of the water main to the affected customers. Only customers on the main that is going to be shutdown actually get this letter.
- On site customer notification – This is probably the most important communication element of the project. It is this opportunity that the contractor advises the customer their service will be interrupted for about 20 minutes and if needed the contractor can coordinate a little delay in the service switch.
- Mobile signage – Mobile signs are placed in front of trenches with project title, council member, contact information, etc…
- Bi-weekly meetings – These Meetings are between MWS and contractors to discuss customer concerns and project progress.
- Customer service hot line – A hotline, answered 24 hours a day, 7 days a week, is provided to the affected customers.

Typical Project Construction

Within Nashville's WIR program, cleaning and lining main rehabilitation is deemed appropriate when there are no structural defects with the mains, the mains were installed (bedded) properly, and there is not a need for a larger main due to current or future growth of an area. A typical candidate for cleaning and lining is a main with few breaks or leaks that exhibits poor flow qualities and poor water quality due to corrosion deposits.

The cleaning and lining process consists of mechanically scraping the inside of the pipe with a drag scraper to remove all the corrosive deposits. Once the pipe is sufficiently cleaned (scraped to bare metal) and sufficiently de-watered, a thin lining of cement mortar is centrifugally applied to the pipe then smoothed with a trowel. The service lines to each home or business are cleared before the mortar hardens.

During the time period a main is out of service for the cleaning and lining process, temporary service lines are installed on the ground surface to provide water to all customers of the affected area. The temporary supply lines are put in place with jumper lines to outside water meter boxes. This temporary setup may remain in service for up to 3 weeks. Meters are bypassed and bills are estimated for this period of time. In some instances temporary fire hookups are also placed in the temporary waterline so that water would be available at that certain location in the event of a fire.

MWS workers precede the rehabilitation effort and replace service lines from the main to the meter, fire hydrant leads, and fire hydrants for the affected areas. Any valves on the main to be rehabilitated are also replaced with new valves. In earlier projects MWS did not replace the services and the fire hydrants or hydrant leads and determined that the further away from the cleaned and lined main, the less benefit realized from the rehabilitation.

Excavations in the streets are limited to access pits for the cleaning and lining equipment. These pits are secured with shoring devices to prevent danger to the public and to the workers. These pits are located where the actual cleaning and lining process takes place and typically at a service line or pipe fitting. Generally there is a pit every 500 feet, wherever there is a significant bend in the line, or where a valve will be replaced or installed.

Study Area Description

Data for this report analysis originated from a combined commercial and residential area. The distribution system within this area consisted of an unlined cast iron gridded network system with relatively few dead ends. Data was obtained from 11 separate cleaning and lining runs for a total 11,350 linear feet of lining. All 11 segments were 12-inch diameter cast iron. During the rehabilitation, water service was maintained to 252-service connections by placing 26,700 linear feet of 2-inch and 4-inch diameter PVC temporary pipe.

Task Rate Comparison

Figures 4.33 through 4.36 display Nashville's pipe rehabilitation program as compared to other utilities participating in the study. The data are segregated by lining and temporary service tasks and also by man hours and days per task. The solid line running horizontally across the figure, represents Nashville's data. The background box plot represents the average utility data as analyzed and presented in Chapter 3.

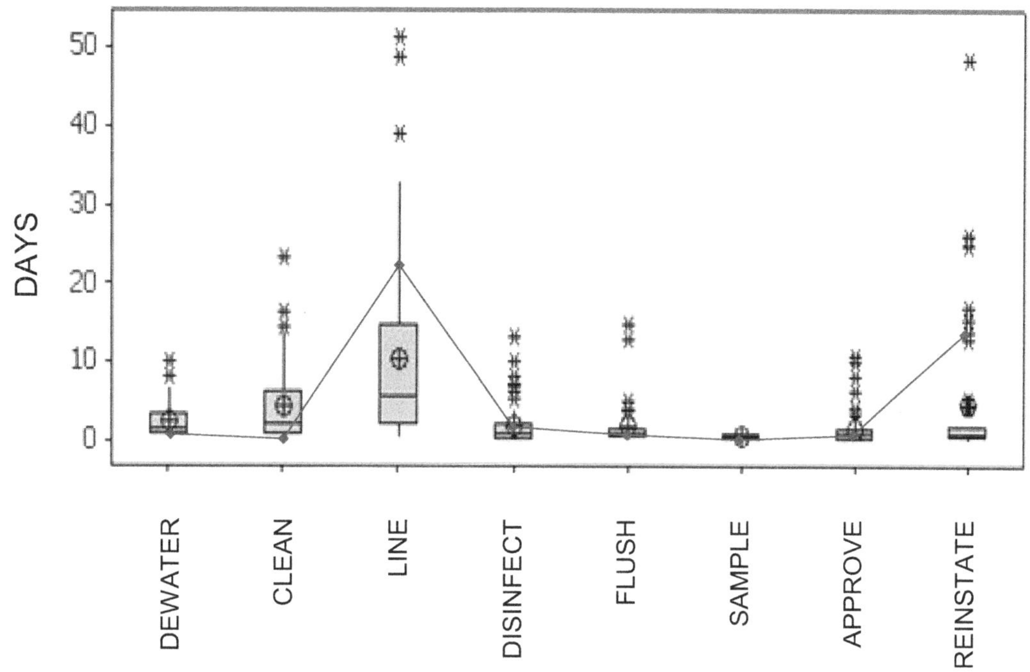

Figure 4.33: Comparison of Nashville's cleaning and lining operation by work day

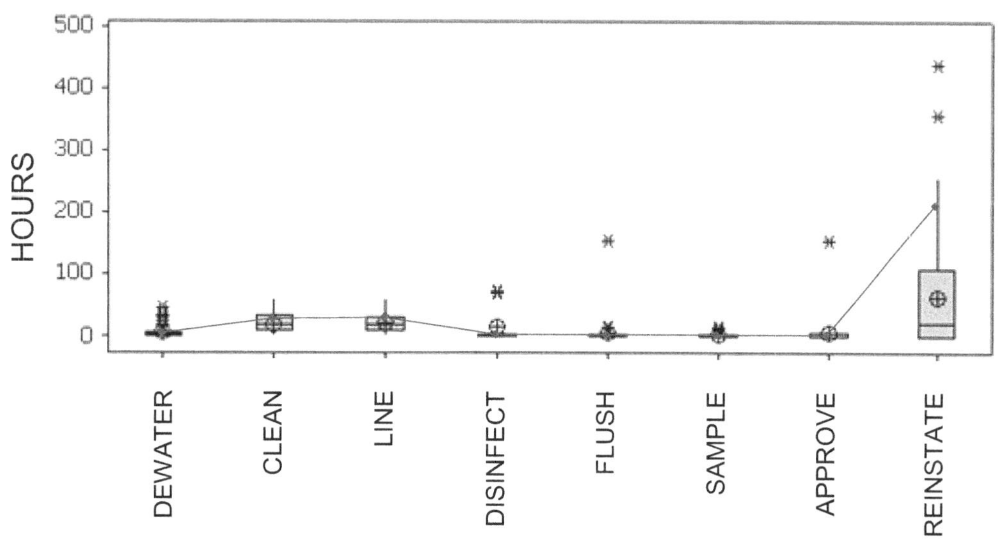

Figure 4.34: Comparison of Nashville's cleaning and lining operation by man hour

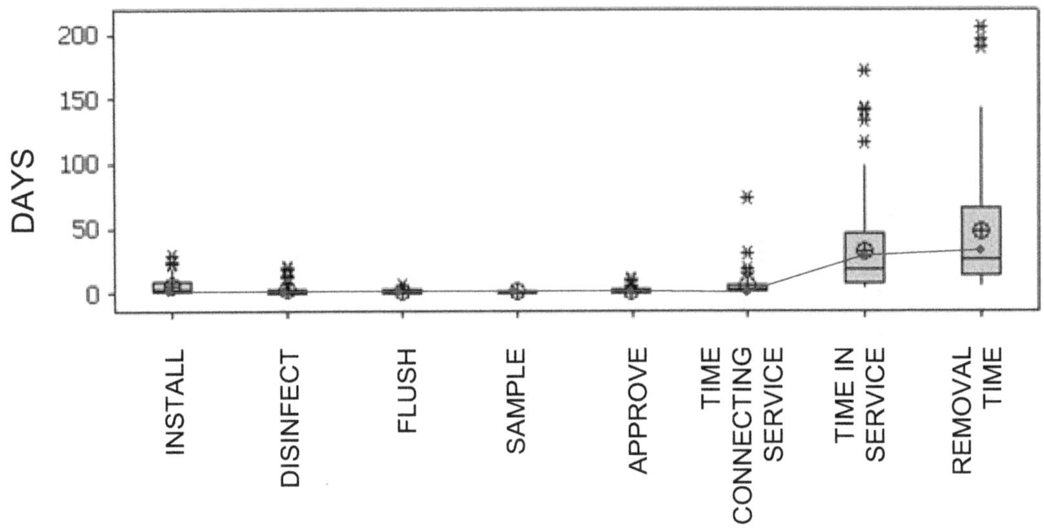

Figure 4.35: Comparison of Nashville's temporary water supply operation by work day

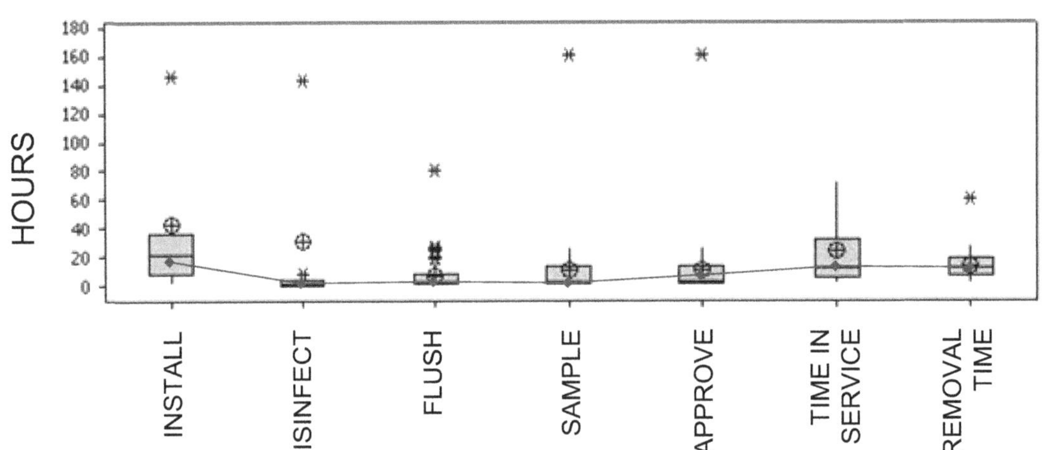

Figure 4.36: Comparison of Nashville's temporary water supply operation by man hour

CHAPTER 5
BEST MANAGEMENT PRACTICES

INTRODUCTION

The purpose of the best management practices chapter is to offer program and project manager's insight into the routine tasks that compose a water pipe renewal program. This section is intended to assist the utility's personnel from the initial stages of program development through inspection of project construction. When applicable, the material is presented in a "checklist" and "commentary" format. The purpose of the checklist is to provide a quick overview of critical items essential to successful project completion. The commentary is designed to provide a more detailed discussion and to present options and alternative procedures. Much of the information contained herein was gathered at the March 15, 2005 workshop held at the University of Louisville, Louisville, Kentucky.

In addition to the checklist and commentary, a typical work schedule for a pipe lining project is provided. The schedule was developed using the data gathered from the participants in this study and presents a baseline to estimate manpower and construction duration. By way of the scheduler a utility can visualize the time requirements for each task and the effect that task has in the overall course of the project. By developing a critical path within the project tasks, anticipated lead time can be built into the construction schedule. The schedule provided is not intended to be applicable to all lining tasks or sites. Rather, the scheduler is a guide that can easily be modified to reflect a particular utility's site conditions and setting.

GENERAL PROGRAM DISCUSSION

Program Development – Checklist

Advanced pipeline renewal program planning will significantly affect the day to day operations of the relining work. Time and effort can be lost when directions are misunderstood or contractor disputes occur. Thus, addressing these overall planning issues early in program development will help reduce overall project time. A detailed checklist for program development is not appropriate as task specific items are difficult to define. As such, the checklist presents a general list of items that should be considered as each lining program is developed.

- Gather data
- Define the problem
- Obtain utility management support
- Clearly define purpose of pipeline renewal program
- Determine an effective renewal method
- Set annual agenda
- Complete general specifications
- Design individual project components
- Define contractor objectives
- Bid and award project

- Monitor contractor progress

Program Development - Commentary

Multi-year Program Development Considerations

Getting Started - A good starting point for a utility taking its first steps toward asset management and pipeline rehabilitation is a comprehensive audit of the entire pipeline network. Pipeline attributes such as pipe type, pipe diameter, and year of installation, along with historical performance data such as leaks, breaks, fire flow availability, discolored water issues, and pressure complaints, should all be compiled and entered in an expandable, user-friendly database. A GIS mapping system that compliments the database and has the capacity to accept pipeline attributes is an excellent tool to display pertinent pipe segment characteristics. Information from any maintenance activity on any pipeline should be entered into the database and on the mapping system for future reference. Developing and maintaining this data is critical to a successful program.

Identifying Problem Areas – Leaks, breaks, pressure complaints, and discolored water are all indicators that pipelines are in need of attention. However, with a properly assembled database, utility asset information can help to more precisely define the need and extent of pipe network problems. Analysis of information in the database, relative to specific pipe segments, material types, sizes, manufacture date, performance criteria, or other defining characteristic will likely generate patterns indicating those assets most in need of renewal. As an example, cast iron pipe installed prior to 1928 is generally unlined pit cast. For the most part it has a very heavy wall and in many cases has an excellent service history with regard to leaks and breaks. However, this material is very susceptible to internal corrosion, which in turn creates discolored water and reduces fire flow capabilities. This unlined pit cast pipe is generally considered structurally sound enough to clean and line or coat with a non-structural material. After 1928, however, the cast iron pipe manufacturing processes changed from pit cast to centrifugally cast. Thinner walled pipe was fabricated, and in the earliest years of the process, the molten iron was not always cast symmetrically, resulting in uneven wall thicknesses. Many utilities that closely track leak and break history can easily pinpoint this pipe through maintenance records and break frequency. Generally this pipe is more likely to need replacement or structural rehabilitation than pit cast pipe.

Clearly Defined Program Goals - As a pipeline rehabilitation program is initiated, program goals should be developed, clearly defined, and conveyed to those individuals with a vested interest in program success. It is important that a strategic plan be developed, at least as a guide, to give the program a sense of direction. Whether the program is developed to eliminate discolored water, improve fire flow capabilities, or replace failing pipelines, the intent must be clearly stated. Each utility must determine what issues are most critical and begin to prioritize pipe segments and developing project scopes. As the program matures, and certain critical pipeline issues have been successfully addressed the shift of goals to new priority items should be documented.

Utility Management Support - Buy-in from utility leaders is a critical element of a successful program. Without support from the upper management, capital funding will be in a constant state of flux. Educating the utility leaders, local politicians and general public, of the benefits of having a consistent asset renewal program may be required. Documented program success and benefits from other utilities along with a clear set of program goals should be included in the education process.

Set Annual Program Size – Determining how much of the aging system to address annually is often one of the more challenging issues for a utility as it develops a program. Competition for available capital funds and labor resources generally establish what percent of the system a utility is able to address each year. Utilities commonly use a figure of 1% of the total network annually as an acceptable goal for their renewal program. At that rate a utility can assume it will address all of its system within 100-years. Some Utilities approach pipeline asset management more aggressively and may budget to renew as much as 3% annually knowing that the actual project length will fluctuate depending on site specific conditions.

Select Renewal Methods - An integral part of developing a program to strategically address a utility's pipeline network is to establish a system to identify pipes for lining or alternative rehabilitation. Each utility must determine if they have pipe that is structurally intact and can line or coat the interior wall with a non-structural material or if the pipes have been substantially compromised due to corrosion and should be replaced. In addition to structural integrity considerations, each utility must also determine if non-structural lining will be a cost effective option.

If the program is to include multiple lining methods there must be a clear delineation between which pipe segments are structurally suitable for non-structural lining and which pipe segments must be replaced with a fully structurally component. Generally, utilities will set a maximum number of leaks and/or breaks per unit length and segments that exceed the maximum are replaced; those with fewer are considered candidates for lining. While utilities may vary their threshold for allowable breaks per unit length, it is important that each utility establish an absolute point at which a pipe is considered sound and that at which it is considered to have reached the end of its useful life.

Annual Program Planning

Define Project Areas - By evaluating historical documentation of the pipeline network and following the predetermined performance measurement criteria, a prioritized list of critical segments can be generated. Once identified, project areas can be defined. Large contiguous areas that contain a high concentration of unlined cast iron mains usually offer better opportunities for design and construction efficiencies. Structures such as flood walls, railroads, and interstate highways, where there may be a high incidence of dead end mains are also commonly used as boundaries between project areas. It is best if project areas have at least one side adjacent to a high quality water supplied through either an already lined or PVC pipeline. In this manner, the work can be sequenced to bring "good water with them" and prevent drawing water from an unlined source into newly lined mains.

Spread the Work throughout the System - In order to share the benefits of the renewal program equitably among all the rate payers, most utilities divide their network into areas or zones. These zones are often delineated along pre-existing political, geographic, or pressure zone boundaries. Annually budgeted funds are then distributed throughout the various areas to insure some renewal work is performed in each area on a regular basis.

Big-Picture Approach – With respect to individual neighborhoods, a utility must take a big-picture approach to pipeline network renewal. Not only should it be concerned with the remaining life of existing lines, but also should consider current and future hydraulic needs. Prior to any relining work, it is essential that the current neighborhood needs are appropriately determined and will be met at the conclusion of the project. If a larger pipeline is needed in an area to properly supply customers, it should be included in the rehabilitation scope even though, it is not technically a relining project. This may impact the funds budgeted for pipe rehabilitation, but project construction will inconvenience the customers only once rather than twice.

It is sometimes difficult to predict future neighborhood hydraulic demands due to short term demographic changes. Many municipalities are experiencing a shift in population from the urban areas to outlying suburbia, thus decreasing the demand for water in the existing system while creating a demand elsewhere. Urban renewal projects, however, are designed to revitalize certain areas and bring back people and industry. Thus it is important to coordinate with city planners so that rehabilitation work is in conjunction with overall city objectives.

Coordinate Activities – Contact local fire department officials to alert them to the upcoming project and solicit their input on existing fire fighting capabilities, fire hydrant spacing, etc. Since one of the advantages of a renewed pipeline is greatly improved fire fighting capabilities fire department officials are usually willing to help in any way possible with these projects. Many fire departments are responsible for hydrant flushing programs they may have a wealth of flow data that would be helpful in determining the more critical pipelines to address.

In addition to local fire departments, local public work departments should also be notified of the planned work activities. Few things are as annoying to customers as a utility cut crossing a newly paved street. Thus, it is essential to coordinate activities with other municipal officials to avoid duplicating work activities or interfering with active or recently completed projects.

Neighborhood Demographics – The local demographics of renewal areas are important to consider during project scope development. Commercial and industrial areas may require substantially more fire protection then residential areas. As such, larger diameter temporary pipelines may be necessary and thus installation time and maintenance will be higher. If conditions allow for a project scope to be developed in an area that is predominately commercial or industrial, the utility may want to explore bidding that area independently from residential neighborhood. However; to attain the best possible contract value the area should be somewhat contiguous to prevent the contractor from remobilizing repeatedly.

Contract Issues

Specific Contractor Volumes – When proportioning project work for contractor bidding, the volume of work included will affect the overall project cost. Larger projects generally offer economies of scale and can reduce unit costs significantly, however, they may limit competition and poor performance can be devastating to a long range program. In order to encourage contractor competition and take advantage of economies of scale, many annual pipeline renewal contracts include at least 25,000 linear feet. Much larger scopes, along with continuing service, or renewable contracts have also worked in favor of some larger utilities.

Define Successful Project – Set measurable goals and expectations for the construction phase prior to bidding a project. It is important to clearly define specific contractor performance criteria such as a minimum length of pipe to be lined per week, maximum time temporary pipe can be in place, and maximum number of customer complaints. Once defined, track and manage these issues throughout the project. During the regular contractor meetings discuss performance changes and continually convey the importance of maintaining a high customer satisfaction level.

Contract Estimate – The construction contract should contain a total cost estimate, a payment schedule and unit costs for work items anticipated on the project. Developing a project estimate for final construction can be extremely difficult if the utility's record drawings are not complete. Building in a contingency fund to pay the contractor for small miscellaneous tasks has proven to be an effective way to handle minor change orders for some utilities.

Project Design

Original installation Records Research - Once the basic project area and proposed scope are established, a more detailed research effort may be required to verify the accuracy of the project planning documents. Many of the larger utilities have detailed, accurate laying schedules and drawings of their original pipeline systems; noting pipe lengths and weights and fittings and valve locations. Others may have no records whatsoever and rely on maintenance activities to gather pipeline information. The purpose of this verification process is to locate interconnections or any possible obstructions in the pipeline that may not be shown on the existing pipeline record drawings. Identifying all interconnections will eliminate lost time during the isolation of a work area. Avoiding an obstruction during the construction phase of the project prevents unnecessary change orders and costly down time for the contractor.

Visit the Site - Taking the time during project design to visit the site can offer invaluable insight to the project area. A site visit is a good way to develop an idea of service density, typical meter vault locations, potential traffic control issues, and access excavation locations. Additionally, other utility work that has affected the water line alignment may be in progress or recently completed. The street may have been recently paved or new large diameter services or fire hydrants may have been recently connected to the main. Any one of these issues could be grounds for a change action if left unaddressed until construction.

Develop Plans and Specifications - It is advisable to produce a detailed set of contract documents to clearly define the scope of work and the utility's expectations. Drawings should include all elbows, tees, crosses, offsets, and valves. Generally fittings not shown on contract documents create an obstruction and potentially shutdown of the construction process. The specifications should contain standard language on AWWA standards for in situ cleaning and lining of water mains.

While developing plans, thoroughly evaluate the pipeline and connections in each street. Lead and galvanized steel service renewals, fire hydrant leads, large domestic services, and fire service connections may each require an excavation or access point. If the finished roadway is going to be greatly compromised by numerous excavations an alternative pipe renewal method may be warranted.

Construction

Permits - Many utilities are required to obtain City/County/State roadway permits prior to construction activities. Permits may cover excavations, maintenance of traffic control, and parking meter issues. The lag time between requesting permits and receiving permits should be built into the project manager's schedule to ensure the contractor is not delayed. Additionally, flushing hyper chlorinated, and in some instances potable water, may be an issue for some communities. The large volume of water flushed during pipe renewal projects may require permitting from the local storm water or sanitary waste agencies. Requesting any necessary permits well in advance of work activities is essential to maintaining maximum efficiency during construction.

Plan Ahead - The objectives of pipeline renewal projects are to rehabilitate the pipe as quickly as possible, with little disruption to the consumer, and without compromising water quality. Successfully achieving these objectives requires close coordination of the project manager and construction inspector. They must always be planning for project's next move as permitting, no parking signs, street closures, traffic control plans, water shut offs all require advanced notice. Large projects, running at peak efficiency may have a value to the contractor of $20,000 per day. Thus any avoidable delay becomes very costly.

Public Notice – Early warning of an upcoming project through public notification may save a utility from unnecessary customer complaints. Public notification of work activities helps set the customer tone for the project. Public notification can be accomplished in a variety of ways but generally includes a mailing to affected customers with an explanation of the pipe renewal process, what to expect in the way of inconveniences, and a description of the benefits of the renewal program. In some instances public input may be necessary during the planning and design stages of the project. Some utilities prefer to hold public meetings and announce the upcoming projects in neighborhood newspapers. If a public meeting is to be held it is best to include local politicians, fire department officials, and several utility representatives.

For commercial, industrial and other customers who rely heavily on a consistent water supply, notification may necessitate a visit from either the project manager or construction inspector to discuss potential water outages and arrange work activities. Schools, hospitals, and customers who rely on a continuous flow of water for their operation require a higher level of attention than average. It may be necessary to limit water outages to nights or weekends for some customers.

Site Review - As a standard practice the construction inspector and the contractor should jointly review a new work area immediately preceding work activities. The site review should include a preliminary inspection of all valves that will be used during the project to verify their location and accessibility. The contractor can also locate and mark excavation points and request for other utilities to locate their respective facilities. Critical customers can be identified and decisions made on how best they should be served. The temporary pipeline configuration can be determined as well as special provisions for commercial drives, wheelchair ramps, etc.

Regularly Scheduled Project Progress Meetings - Mandatory project progress meetings are vitally important to the successful management of large scale programs. Field personnel and utility management should be required to attend. Depending on the complexity of the project these may be monthly or even weekly meetings. The project manager should lead these meetings and have a prepared agenda that includes progress to date, upcoming work, unresolved customer issues, etc. Contractors should be given the opportunity to discuss what improvements could be implemented to increase productivity. Open communication between the contractor and the utility is absolutely crucial to the project's success.

Production Monitoring - During the actual cleaning and lining process the construction inspector should be on site observing the operation. Once the pipe is cleaned the inspector should check the thoroughness of the cleaning either visually or using a closed circuit television and remote camera (CCTV). During the lining process the inspector should monitor various aspects of the equipment to verify proper operation. On cement mortar lining projects pump pressures, winch speed, mortar volume and consistency, and lining thickness are critical components to monitor. Epoxy and urethane lining machines document operational conditions on a continuous basis and provide a printout following the completion of each lining run.

Quality Assurance - CCTV is a common method of inspection following the lining process. CCTV can only show that the pipe wall is coated and can not be used as an indicator that the proper thickness was applied. Intermediate inspections along the lining run are the only way to verify that the lining was applied consistently throughout the pipeline. Frequently the contractor will line through fire hydrant connection points, in line tees, or even valves. As these appurtenances are removed and replaced after the lining operation, these points can be used to inspect the lining thickness and quality without the need for an extra excavation. In the absence of existing fittings that are to be replaced it is advisable to select a random location in the lining run and have the contractor remove a short section of main for inspection.

Working with the Contractor

Cooperation is Key - Partnering with the contractor and jointly solving construction related problems is always the best arrangement. Project delays are costly to everybody involved as the contractor loses income and the utility risks losing customer confidence. Veteran contractors have a broad knowledge base and it can be beneficial to encourage them to share ideas of better ways to accomplish the pipeline projects.

Paying the Contractor - For most utilities it is the responsibility of the construction inspector to verify work quantities in order to generate a pay voucher. Develop a pay request procedure and format that is mutually agreeable to both the utility and the contractor, keeping it as simple as possible. Within the bid package, request unit costs for potential extra work items outside the original scope. This can greatly reduce the time lag associated with negotiating for extra payment and potential work stoppage while seeking change order approval.

As with any business the contractor relies on timely payment in order to meet payroll and project material expenses. To speed the pay process, it is best to have the inspector verify quantities prior to the contractor's invoice submittal. If the work has been completed, as specified, the project manager should expedite payment to the contractor. If the contractor knows that the utility pays its bills in a timely manner future project costs may be reduced.

Alternatively, future project costs will rise if a contractor has to finance materials and borrow money to for payroll due to cash flow inconveniences and expenses.

***Residential Plumbing* -** The bid documents should clearly state what is expected of the contractor with respect to homeowner or private property issues resulting from water service interruption. Prior to bid, determine who will be responsible for residential plumbing problems and how they will be resolved. It is often necessary to clean faucet aerators and to replace flush valves during lining operations. If the contractor is responsible for residential plumbing problems, typical repair expectations and costs should be included in the specifications. Only licensed plumbers may perform residential plumbing work.

TEMPORARY PIPE INSTALLATION

Temporary Pipe Installation Checklist

Typical procedures, methods and considerations for installing a temporary water supply line are as follows:

- Flush all hydrants that are to be used to supply temporary pipe
- Select most practical location for temporary pipe
- Select best material for site conditions
- Install temporary pipe, ramp residential drives and bury pipeline at street crossings
- Install backflow prevention device at hydrant if required
- Connect pipeline to hydrant
- Fill line chlorinate, flush, and sample
- Transfer customers to temporary pipe
- Remove temporary pipe at end of project

Temporary Pipe Installation Commentary

General Procedure

Generally, temporary water lines are installed above ground on both sides of the street in either the gutter or adjacent grass areas. The temporary pipes connect to fire hydrants outside of the project limits and provide continuous water service to affected customers. Rubber, steel or asphalt cold patch ramps are placed at driveways crossing to protect the temporary mains from traffic. The temporary pipes are disinfected to meet bacteriological standards and then customer taps are connected using meter setting or hose bib connections. With this process, customers are without water for only the length of time it takes to transfer them from the main to the temporary supply. However, they will have reduced service the entire time the temporary network is in place.

Layout Considerations

Temporary water supply mains must be installed in the most logical location with respect to water meter location, pedestrian traffic, and vehicular traffic. The temporary pipe and service hoses should cross sidewalks and streets as seldom as possible. Each crossing represents an opportunity for a pedestrian accident or a vehicle hazard.

Depending on the traffic load and driveway elevations it may be necessary to install the pipe along the base of the curb in the gutter. Care must be exercised in these instances to protect the temporary line from vehicular traffic. Often, this will involve either switching to a steel material or saw cutting surface materials to bury the pipe below grade at driveway entrances.

The temporary water system must adequately provide for consumer demand and fire protection. The contractor and the utility should jointly determine likely flow demand requirements and size it the temporary system accordingly. With sufficient fire department coordination and fire hydrant facilities, some utilities have been able to avoid providing fire protection to areas on temporary supply.

It is also important to install an ample number of valves to facilitate maintenance operations and minimize customer outages during repairs. Designing and constructing a well looped or gridded, properly sized system will help eliminate costly, time-consuming customer complaints and potential construction delays.

Material Selection

A variety of acceptable materials are used to provide temporary service during rehabilitation projects. Among those, the three that are more typically used are steel, PVC, and HDPE. Each material has its benefits and its drawbacks. As with any material used to convey potable water all temporary pipeline components must bear the NSF and FDA approvals.

Steel, while being very durable is fairly heavy and more difficult to handle than plastic materials. The bolts attaching the victaulic fittings to each pipe segment present a puncture hazard to vehicles tires. Great care must be exercised to keep the threaded ends of the bolts turned down or toward the curbing to prevent tire damage.

PVC, available in 20-foot lengths, is light weight and easily assembled and disassembled. While the material is easy to work with it is not as durable as steel or HDPE. As compared to other temporary pipe materials, PVC tends to be more brittle and susceptible to impact damage.

HDPE has proven to be very durable pipe material if installed appropriately. Small diameter material can be delivered to the site in 500-foot long continuous spools or in straight lengths as long as 40-feet each. Joining HDPE pipe is typically accomplished using butt-fusion welds which makes disassembly and reassembly slightly more cumbersome than PVC and steel materials. As with other pipe materials, the HDPE can be reused in subsequent areas if the pipe is cut into manageable segments, transferred to the new site, and rewelded.

Nitrile synthetic rubber hose material, while not common in the U.S. was used very effectively by one of the study's participants, NKWD. This material is manufactured in 50-foot lengths, coupled together, and stored on spools of approximately 500-feet. The material is very quickly deployed on the site and requires no special ramps at residential drives. However, service connections can be made only at the hose couplings which may create longer lengths of individual service line hoses.

Water Supply Points

Generally the temporary pipeline is connected to fire hydrants located on streets adjacent to construction areas. In the absence of reasonably accessible fire hydrants, taps can be installed on adjacent mains to supply the temporary line. In either case, it is important to verify the selected main will remain active throughout the pipe lining operation.

Prior to attaching the temporary pipe to an existing main or fire hydrant the feed system should be flushed to assure the water feeding the temporary pipeline is of the best possible quality. Fire hydrants should be inspected for leaks at the base, and sounded to verify drain holes at the hydrant elbow are not allowing water to escape into the subsurface.

Meter Location

In warm climates meters are easily accessed in pits near the property line or right-of-way. The temporary supply may connect to the inlet side of the meter, or in the case of compression type setters, may replace the meter with a straight pipe and tee assembly. If the temporary supply can be connected to the meter inlet, it is still possible to meter customer usage during the pipe lining activities.

In northern, colder climates, meters are more frequently installed inside homes and businesses to prevent freezing. This location presents a number of challenges for the utility and the contractor. If available, the temporary service can be attached to an outside spigot. In the absence of an outside spigot it may be necessary to gain access to the inside of the structure and effect the connection at the meter. A third alternative connection point is at the curb stop, but this requires an expensive and time consuming excavation. The curb stop location, however, creates a positive air gap between the pipeline construction and the temporary water supply. The outside spigot connection and the inside hook-up both depend on the curb stop to provide a positive shut off.

Domestic Service Connections

Residential service connections are typically made with 3/4-inch or 1-inch flexible materials. Multi-unit structured will likely require larger size hose connections depending on the number units being served. As these materials convey drinking water, they must be NSF and FDA approved. Additionally, during installation and throughout the project it is important to maintain the materials in clean, water tight working condition.

Occasionally, a residential customer will have a larger than average service tap. Many older homes were plumbed with toilets that are fitted with flush valves that rely on flow and pressure to shut off properly after operation. These homes may require a larger diameter supply in order for the plumbing fixtures to function properly. The diameter of the meter is often used to determine the size temporary service required. Locate and note any large diameter service requirement and determine an appropriate supply strategy.

Hospitals, schools, child care facilities, kidney dialysis centers, etc rely on a constant uninterrupted supply of potable water in order to remain operational. These customers require a higher level of attention than most residential customers. In order to provide the appropriate level of customer service these situations will require the personal attention of the utility representative as well as the contractor.

Fire Suppression Systems

Maintaining temporary supply capacity for consumers with individual fire suppression systems presents a number of challenges. It is advisable to seek input and approval from the consumer, local fire protection districts, and alarm monitoring companies before a temporary supply system is installed. Additionally, it is advisable to thoroughly inspect the customer's fire suppression piping prior to transferring to temporary supply. The customer's internal plumbing may not be in good repair and pressure fluctuations during the transfer and service work may create leaks inside the structure. It may be necessary to isolate the fire suppression system from the existing water supply to prevent backflow of antifreeze or degraded water.

A simple method to provide continuous supply to a consumer's fire suppression system is a direct feed to a Siamese hydrant fitting. While this will provide adequate capacity, it may interfere with the local fire district's ability to properly attach to a hydrant.

An alternative, but more reliable method to provide temporary fire suppression supply is to excavate the customer's service line near the right-of-way and make a connection directly to the incoming pipe. This method requires coordination with the customer to determine an acceptable time to have the fire protection service off line for the transfer to the temporary pipe. Care must be exercised to assure the connection to the customer's line is restrained in some manner to prevent the connection for separating. Backflow prevention devices should be used to prevent counter flow from the sprinkler line.

A third method of supplying fire suppression capacity takes advantage of the excavation made at the junction of the buried service line and the primary water main. By leaving the tee or tap connection in place and making a cut and cap on the main directly adjacent to the tee, the fire service can be fed from one direction of the water system. Once the dewatered segment is cleaned, lined, and ready to be returned to service the existing service tee and valve can be replaced with new appurtenances and connected to the newly lined pipe segment. The tee is then plugged on the opposite side and the second segment is cleaned and lined. The plug is removed and final connections made subsequent to approval of the second segment. This process may require more up-front planning than the others, but no temporary pipe is installed and the customer is provided a full fire suppression capacity for the project duration.

Elevated Pressure Zones

Most utilities maintain multiple pressure zones throughout their system. In some jurisdictions the pressure within a specific zone may exceed 100-psi. Excessive pressure can pose a significant hazard in the event of a temporary pipeline or service hose break or fitting failure. In extreme cases it may be necessary to install pressure reducing valves (PRV's) at the hydrant or other supply point in order to reduce the pressure within the temporary supply network to a safe limit.

Vehicular Crossings

PVC, HDPE, steel and large diameter nitrile rubber hose pipe all require some form of protection to enable customers to cross the pipe without damaging it, or their vehicles. Common protection methods include ramps made of cold patch bituminous asphalt, dense grade aggregate, or a mixture of the two. While these ramp materials provide adequate protection, they require routine maintenance and create customer interaction problems. The gravel ramps must be continually replaced as the material migrates, and the asphalt ramps stain sidewalks and driveways. Often these materials are tracked into customers' homes and places of business and are the source of complaints.

As an alternative to asphalt or gravel, lumber is less common, but equally effective ramping method. A benefit of a lumber ramping system is that any configuration can be constructed by on site personnel. This is particularly useful when incorporating American Disabilities Act (ADA) access mandates with allowable slopes and handrail specification. While very adaptable, lumber is less durable than other products and requires an increased level of labor to construct.

DISINFECTION PROCEDURES

Disinfection Procedure Checklist

Typical procedures for disinfecting a newly line main or temporary water supply main are listed below:

- Verify curb stops and connecting valves are in the closed position
- Verify air release valves are in the open position
- Verify plug ends are properly blocked and braced
- Use backflow prevention device on liquid chlorine systems
- Fill line slowly, allowing air to escape as the line is filled
- Test water to verify chlorine concentration, document
- Close fill valve immediately after line is completely full
- Close air release assemblies after closing fill valve
- Allow required contact time
- Connect de-chlorination equipment if required
- Open flushing connection before opening fill valve
- Open fill valve
- Test water to verify chlorine concentration, document
- Flush long enough to change water over several times
- Test water to verify chlorine concentration is that of typical finished water
- Disassemble de-chlorination equipment if used
- If practicable open hydrants and fill valve fully to induce a scouring velocity flush
- Reduce flow velocity
- Flush until clear
- Collect water sample for laboratory analysis
- Close fill valve prior to closing flushing connection
- Close flushing connection

Disinfection Procedure Commentary

General Requirements

The Safe Drinking Water Act (SDWA) requirements governing normal operation of a water distribution systems are equally enforceable during maintenance activities and when providing temporary water service. Typically, after any repair or modification to the distribution system, the affected components are flushed and disinfected according to the AWWA Standard for Disinfecting Water Mains (ANSI/AWWA C651-99) before being returned to service.

Flushing

Preliminary flushing is a straightforward procedure by which large volumes of water are directed through the pipe and discharged on the ground or into a sewer system. The purpose of this activity is to remove any dirt or debris that may have accumulated in the pipe and prevent it from entering the distribution system. While flushing velocities are required to be greater than 2.5 ft/sec, there are no requirements for the duration of flushing or required volume. However, flushing is usually continued at least until clear water discharges from the pipe. It is noted that special provisions are made for large diameter pipes where flushing would produce a significant volume of discharged water. Dechlorination procedures must be followed where required.

Chlorination

C651-99 provides three alternative methods for applying a chlorine disinfectant. It is noted that all methods are not equally applicable to all field conditions. In general, the disinfectant method and corresponding contact time as dictated by C651-99 are shown in Table 5.1.

Bacteriological Testing

A bacteriological test sample is obtained once the main has been disinfected, dechlorinated, and flushed. Bacteriological testing generally requires 16 to 24 hours before results are determined. If the bacteriological test passes, the main can be returned to service. Thus, total time after a main has been lined may be in excess of 48-hours, depending on the chlorination methods and bacteriological test procedures implemented.

Table 5.1
Summary of ANSI/AWWA C651-99 Disinfection Methods

Disinfection Method	Initial Concentration	Contact Time	Final Concentration
Tablet method	25 mg/L	24 hours	detectable residual
Continuous-feed method	25 mg/L	24 hours	10 mg/L
Slug method	100 mg/L	3 hours	50 mg/L

GENERAL CLEANING AND LINING PROCEDURES

Cleaning and Lining Checklist

A typical main cleaning and lining process involving the equipment and work associated with providing temporary service are listed below.

- Confirm all service connections have been relocated to the temporary pipeline
- Close line valves to totally isolate section of main to be cleaned

- Verify segment isolation by opening a service line or fire hydrant before cutting pipe
- Remove short pieces of the main at required intervals to clean pipe
- Clean pipe with an approved method, remove excess water
- Inspect line with CCTV if required to verify removal of corrosion deposits and that service connections are not leaking into the main
- Apply approved lining or coating
- Inspect lining following product manufacturer's recommended cure time
- Replace pipeline appurtenances
- Fill and chlorinate using approved method
- Flush line following appropriate chlorine contact time
- Collect water sample for bacteriological laboratory testing
- Reinstate main and reconnected service lines upon receipt of satisfactory bacteriological test results

General Cleaning and Lining Commentary

General

The pipe cleaning process serves to remove internal corrosion, reestablish the original inside diameter and to create a clean working surface for lining application. The lining works to provide a protective barrier to resist future corrosion. The cleaning process is most often accomplished using steel drag scrapers pulled with cables or by use of a rack feed bore machine. Linings are typically either a cement mortar mix, a two part epoxy resin, or a two part modified polyurea.

Access Points

Access pits for the cleaning and lining operation can be identified and excavated as the temporary water supply pipeline is being installed. Typically, these points will be at existing valves, tees, crosses, bends, and fire hydrants.

Existing Valves and Tees

The cleaning and lining process can accommodate some minor deviations during a run. It is not absolutely necessary to remove gate valves along the segment being cleaned and lined, but verification that they are fully open is essential. Butterfly valves, however, must be removed to allow the cleaning and lining equipment to pass through the pipeline. It is possible to line through some bends, but close inspection of the final product is required to insure the coating is evenly distributed and that the lining equipment itself did not damage the coating. Lining through tees, crosses, and gate valves prior to their replacement may enable the contractor to line longer lengths of pipe from a single set-up, thus improving production. Lining through appurtenances that are subsequently replaced provide the opportunity to verify lining thickness and quality at mid points.

Keep Fire Department Informed

During the lining operation, it is important to maintain communication with the local fire department. Daily notification to fire department representatives of fire hydrants and mains returned to service and those removed is typical. Additionally, fire departments should be aware that mains in the final stages of disinfection may be used during an emergency even though they have not been returned to full service. Routine visits to the local firehouse to keep the department informed will serve to reinforce the water main rehabilitation program's commitment to improve the safety of the community.

Keep the Project Moving

For a well planned, scheduled and operated cleaning and lining operation, a variety of tasks are performed simultaneously. As cleaning and lining is performed in one location, the temporary pipe is taken down from the previous work area and set up in the subsequent work area.. Only during start-up or as the project comes to close are some tasks not continuous. For a typical project, temporary pipe may be installed and removed from different sections nearly every day; similarly, existing pipe may be removed or returned to service. This rapid sequencing of events requires a high level of planning and scheduling. The customer will benefit from the rapid pace as this will reduce the length of time the temporary supply is required.

Sampling Water is Time Critical

Timely collection and processing of water samples is essential to keep the project on schedule. Temporary pipeline water samples are required each time it is dewatered and reinstalled. These samples typically take 24-hours to process in the laboratory, thus a delay collecting samples can significantly affect the production schedule. When accounting for the number of temporary pipeline segments installed during a project, efficient collection and processing of samples is critical.

Lining Inspection

During the lining run, performance data can be used to judge the effectiveness of the lining operation. Lining operators can adjust the rate the machine is pulled through the pipe and the pressure the lining material is pumped into the lining machine. Both of these elements should be monitored throughout the lining process and compared to pipe diameter and lining thickness. A quick calculation to determine the amount of material applied per foot by inch-diameter can provide a rough check on contractor operations. If major irregularities are detected the lining operation should be stopped and the lining material immediately removed if possible.

While it is not possible to actually watch the lining machine during the lining process, it is usually possible to hear it. Although not an exact science, it is possible to detect variations in the compressed air driven machine tone during operation that may indicate if it is functioning improperly. As a quality assurance check, epoxy resin and polyurethane lining equipment offer a computer printout which indicates material ratio, volume, and rate of travel during the lining process.

Inspection of the applied lining is an important safeguard in the rehabilitation process. However, the lining or coating can not be easily observed as it is being applied, thus inspectors must rely on a combination of limited visual inspection of the finished product and performance data. The best opportunity to visually inspect the lining is as the machine is inserted into the pipe and at the end of the run when the machine is removed. The inspector should verify that the coating meets the specifications with respect to thickness and appearance. Flashlights or mirrors can be used to visually inspect the surface quality for limited lengths inside the pipe. An inspection of the entire pipe length will require use of closed circuit television (CCTV).

Directional Flush During Reinstatement

While newly lined mains are flushed and sampled prior to returning to service, there is a possibility of corrosion debris accumulating against the closed connecting valves. To prevent this material from entering the water system network, directional flushing immediately prior to reinstatement may be necessary. With a fire hydrant opened on the newly lined segment, connecting valves should be opened one at a time to allow water to flow into the newly lined pipe and be discharged from the hydrant. After thoroughly flushing the opened valve should be closed and the next sequential valve opened. Directional flushing of the newly lined main greatly reduces discolored water resulting from valve operation.

Customer Relations

Customer complaints occur most often during the construction phase of the lining operation. During this time, the contractor normally has numerous on site personnel and there is a lot of activity. Additionally, many of the contractor operations are time critical and there is a need to expedite tasks. If a customer complains during this time, every effort must be made by everyone with a vested interest in the project to address the issues in a timely manner.

CUSTOMER RELATION CONSIDERATIONS

Customer Relations Checklist

Listed below are some general items to consider when interacting with customers.

- Contact customers prior to beginning construction
- Issue utility identification badges to contractor personnel
- Contact customer prior to turning water off to transfer service from existing main to temporary pipeline
- Designate individual to be responsible for 24/7 customer response
- Respond immediately to customer calls
- Monitor condition of temporary pipe daily; repair deficiencies immediately
- Adhere to work hours, start and stop times, agreed to in notification
- Do not block driveways with equipment
- If equipment must block a driveway notify customer in advance
- Supply temporary toilets on site
- Do not litter or leave construction debris on site
- Obey neighborhood speed limits
- Refrain from foul language and cursing

Customer Relations Commentary

Complaint Resolution Process

In order to maintain consistency, one person within the utility, professionally trained in customer service, should field the calls from customers during regular working hours. When a complaint is received, the construction inspector should respond to the customer and make every effort to resolve the issue. The inspector should be empowered to make decisions and work with the contractor to resolve complaints as quickly as possible. As a clause in the contract, some utilities require the contractor to respond to customer generated complaint issues within a designated timeframe. If complaints are not resolved within a timely manner, monetary penalties may be assessed. Only in extreme cases should the contractor be sent alone to meet with a customer who has registered a complaint with the utility.

Complaint Call-Center

The project manager should routinely verify that call-center employees are aware of the pipeline renewal projects and their geographical limits. A map showing affected streets is an effective way to convey frequently changing information. A complete listing of the streets and the limits can also be helpful for call-center personnel. The call-center staff should be educated to a basic level on the pipeline renewal process to better serve the customer. If possible, call-center personnel should be provided the opportunity to visit the site during construction so as to gain first hand knowledge of the process.

Professional Plumbing Services

Neither the inspector nor the contractor should enter a residence or business establishment and attempt to provide any plumbing repair. In most jurisdictions, only licensed plumbers may provide repair services on the private side of the service line. If plumbing repairs are necessary, either the utility or the contractor should hire a licensed plumber to complete the work. A plumbing contract should be arranged in advance of the cleaning and lining construction.

Customer Interaction

The general public sees the contractor as an extension of the utility. It is important that the contractor's workforce fully understand that they must treat all customers with the respect at all times. This includes, but is certainly not limited to: using appropriate language, using proper toilet facilities, obeying traffic laws, refrain from creating excessive unnecessary noise, not littering, parking in appropriate locations, and not blocking driveways unnecessarily.

MS PROJECT MANAGEMENT TOOL (SCHEDULER)

Using Microsoft Project software, Figure 5.1 illustrates the number of days required to complete individual tasks associated with, installing a temporary supply pipeline and cleaning and lining a segment of existing pipe. The initial section of the scheduler, titled *pre-construction* has been included as an aide for project managers in order to visualize the required lead time prior to beginning the actual construction process.

As was done previously in this report, the data has been normalized to 500-feet for the existing pipe segment and 1,000-feet for the temporary pipeline. This enabled a more accurate averaging of the information supplied by the participants. Every effort was made to match the scheduler dates with the participants' data. The scheduler is based on workdays.

The pre-construction task information shown is a combination of information gathered at the workshop held with the participants and the researchers' background knowledge. This list is intended to assist a new project manager by providing reasonable guidelines and a listing of potential tasks items of a successful project. For the seasoned project manager the list will simply serve as a reminder of the tasks order and expected durations.

The six primary tasks associated with placing temporary pipeline in service require a total of nine days to accomplish. Many of these tasks are repeated with each disassembly and reassembly of the pipeline.

The seven tasks necessary to complete the cleaning and lining of an existing main segment required a total of nearly 15 days. Of these defined tasks, lining required the greatest length of time to complete, 5.9-days. One possible explanation for this seemingly excessive amount of time may be the fact that some contractors prefer to clean a sizable amount of pipe before lining any. This method would tend to increase the average number of days between cleaning and lining application for the study group.

The project scheduler represents the tasks and time durations for lining one 500-foot segment of existing pipeline and the corresponding 1,000 feet of temporary water supply pipe.

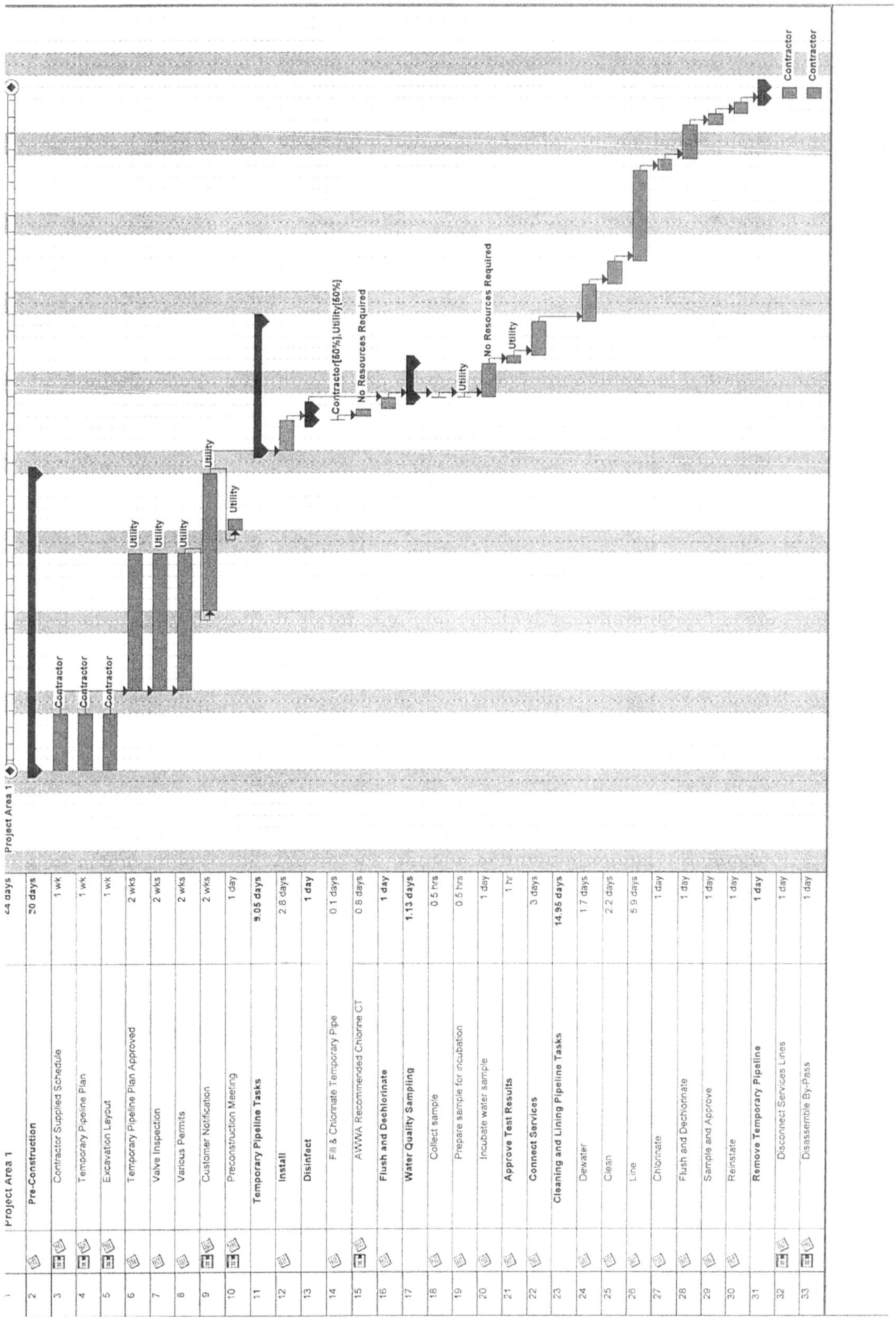

Figure 5.1: Scheduler template for cleaning and lining projects

CHAPTER 6
CONCLUSIONS AND RECOMMENDATIONS

CONCLUSIONS

Maintaining water distribution system integrity and increasing customer satisfaction are two critical, intertwined components of water system operations. However, during rehabilitation operations, these two goals are often at odds. Maintenance operations typically require a service disruption and may include a significant amount of equipment or personnel in front of a customer's house. This is in sharp contrast to the consistent, almost invisible service that water system utilities normally provide to their customers. However, as the water utility infrastructure ages, more maintenance operations and system improvements will be required to insure continued performance. While the infrastructure improvements are ultimately to the long-term benefit of the customer, the short term conflict between the maintenance/renewal operation inconvenience and customer satisfaction will continue.

When repair or rehabilitation of existing infrastructure is necessary, many utilities employ techniques which reuse the existing conduit to reduce the impact to existing surface facilities and traffic flow. While these pipe renewal operations are generally less expensive and less disruptive than open trench replacement, customers must still endure service disruptions, construction equipment and general construction environment.

To reduce the amount of time that customers are physically without water service, pipe renewal operations of this nature, such as cement mortar lining, epoxy lining, main bursting, and sliplining, generally establish an above ground temporary pipe network. While the temporary pipe network generally provides practically uninterrupted service to customers, it still may generate customer complaints and is expensive to provide. In addition, the temporary pipe also represents substantial risk to the utility with respect to system security and performance.

The data collected from select utilities during this project suggest that a number of best management practices have the potential to reduce the length of time necessary for pipe lining operations and consequently, the length of time a temporary pipeline is in service. Each recommended management practice would require a shift or change in what is currently considered a typical construction practice. As the extent of change increases, so do the level of potential benefits. As an example, some minor changes may impact the number of days the temporary pipe is in place, but not impact the man hours required to install and maintain it. Changes of a more sweeping nature might impact both the number of days a customer could expect to have construction on their street and the amount of labor necessary to complete the pipeline rehabilitation. As always, change comes with some associated risk. What follows is a discussion of changes in the current construction processes that have the potential, if properly implemented, to greatly reduce the number of days most customers are subjected to maintenance activities and the temporary pipeline.

RECOMMENDATIONS

Minor Modifications

With relatively minor changes to an existing lining operation, utilities can effectively reduce the amount of time associated with pipe lining activities and ultimately the amount of time that customers are inconvenienced. Subsequently discussed are several options that various utilities employ within their lining operations. If implemented separately, these minor recommendations are anticipated to improve efficiency and reduce downtime by several hours to one day.

Improve Task Sequencing

Currently a number of utilities view a season's pipe lining activities as one large construction project. As such, the utilities do not routinely get involved with scheduling day-to-day operations and typically let the contractor dictate the schedule or work activities. While this approach may lead to efficient contractor operations, it may not provide for the same levels of customer satisfaction, reduced risk (in terms of the amount of time temporary supply is necessary) as other scheduling procedures. As an example, contractors may elect to install temporary pipeline networks throughout the neighborhoods before beginning any actual cleaning and lining tasks. Thus, if a season's contract is for 10,000-linear feet of pipe lining, the contractor's first step is likely to install approximately 20,000-linear feet (both sides of the street) of temporary pipe. The result is that the first piece of temporary pipe installed may remain in front of the customer's residence for several months before the line is assembled, put into service, and then removed after the cleaning and lining operation. While the entire 10,000 feet may have been cleaned and lined in a reasonable amount of time, some of the first customers affected have endured the presence of temporary pipe the entire time the project has been under way, even though most construction activity occurred at other locations within the neighborhood. In many cases this scenario is unnecessary and can be changed simply by changing the construction strategy.

An alternative to viewing a season's pipe lining activities as one large construction project is to reduce the working area footprint and to consider the work as a series of small sequential projects. If temporary pipeline installation tasks and cleaning and lining tasks are sequenced more on a block-by-block basis, the number of days temporary pipe is in place in any one location can be significantly reduced. By working in a small footprint, temporary pipeline can be installed in shorter increments and placed into service immediately after the water has been approved. Cleaning and lining tasks can then begin immediately following approval of the first few segments of temporary pipeline. Subsequently, the temporary pipe can be removed once the newly lined pipe is approved for service and connections are reestablished. And, as the footprint for these operations is only one-block, the total amount of time that the temporary pipe and work crews were operating within the area is substantially reduced. From the customers point of view the lining activity is shortened to only the time the construction passes in front of their residence.

The critical component of sequencing the construction of the reduced footprint is to exactly coordinate the final approval of the temporary pipe installation with the initial cleaning and lining tasks. With the reduced footprint methodology, each task follows the preceding task as quickly as possible without delay or lag time. In this manner the first section of temporary pipe installed is also the first section removed. As the lining process proceeds through the neighborhood, it is very likely that many customers will have the work completed and temporary pipe removed before customers later in the sequence are even placed onto the temporary water supply. This reduced footprint construction sequencing does require an added level of attention to project planning and a greater level of scheduling manpower, however, it also has the potential to significantly reduce the amount of time customers are exposed to construction activities and placed on temporary service.

Change Lining Material

The selection of appropriate lining method or material can significantly effect the duration of a pipe lining project. Within the United States, the two more commonly applied products are cement mortar and epoxy. Cement mortar requires approximately 12-hours and epoxy requires approximately 16-hours to cure prior to reintroducing water into the pipe. As such, the pipes are typically filled and hyper-chlorinated the day after the actual lining was completed. The application of a rapid set material, such as a polyurethane (PU), would allow for the water to be reintroduced into the pipe 1-hour after lining has been completed. Thus, switching to a rapid set material would move forward the disinfection by at least 11 hours and in many cases 1-calendar day. All subsequent tasks would then occur sooner including reinstating the main and removing the temporary pipeline. While switching to a rapid set material would compress the overall schedule, it would not necessarily reduce manpower requirements as each individual task must still be completed.

Change Temporary Pipe Material

The data collected indicated that there is a substantial variation in installation rates for the three temporary pipeline materials commonly used. Of the products assessed, nitrile rubber was documented to require the least amount of manpower to install at a rate of 3.3-man hours/1,000 linear foot. This rate is about 6½ -times more efficient than HDPE pipe and 9½-times more efficient than the PVC installation rate. The rubber material, however, was observed to have a break frequency of 0.74 breaks/1,000 feet installed, about an 85% higher frequency than HDPE, but about 65% less than the PVC. This apparent high failure rate (as compared to HDPE) may be attributed to the limited amount of available data as only one utility supplied data on nitrile rubber temporary pipe. It is likely; however, that the increased labor to repair leaks and breaks in the temporary line and associated customer complaints would offset the labor saved using this material. The advantages and risks of using nitrile rubber for temporary pipe line should be evaluated by the utility on a case by case basis. If nitrile rubber is a feasible alternative, project time could be reduced by several hours per segment.

Moderate Modifications

Utilities willing to reevaluate their operating philosophy and to actively develop new equipment have the potential to further improve the efficiency of their lining operations. Listed below are several recommendations that have the potential to improve performance by one to two days per segment.

Rapid Chlorination

Currently, the AWWA Standard for Disinfecting Water Mains (AWWA C651-99) specifies the tablet method, continuous feed method and the slug method as acceptable disinfecting procedures for "mains that have been removed from service for planned repairs or for maintenance that exposes them to contamination". While AWWA C651-99 allows for three alternatives, most utilities prefer to disinfect newly lined mains with either the tablet or continuous feed methods which require a 24-hour contact time. In addition, many utilities treat each set up of the temporary pipeline as newly installed pipe and repeatedly subject it to the same disinfection routine (24-hour contact time). Considering the number of times the temporary pipe is dewatered, moved and sampled when the lining operation is subdivided into block by block segments, this method of disinfection can greatly increase the number of days temporary pipe is in place in any particular location.

As discussed, most utilities currently prefer to use a disinfectant procedure that maintains specified chlorine residual for a 24-hour contact time. The slug method (AWWA C651-99, Sec. 4.4.4), however, allows for an optional 100 mg/l initial concentration with a 3-hour contact time. Employing this allowable slug method could significantly reduce downtime during lining operations. If used for both the temporary pipe and the newly lined main the savings would be twofold and approximately 2 days could be removed from the process. While the rapid chlorination would likely not reduce the manpower required to complete the project, the number of days the temporary pipe is in place, and thus the associated risks would be reduced.

Pressurized Hose Reel

A second alternative, which would eliminate the need for chlorination at every reinstallation of temporary pipe, is a hose reel designed to hold enough temporary pipe to span at least one standard city block. This would require a temporary pipe material flexible enough to be coiled onto the reel while full of water. The reel itself would need to be substantial enough to hold the weight of the hose and the water contained within. This would allow the temporary pipe to be disconnected from customer services and rolled onto the reel and transported to another nearby location without being depressurized. Service connections would need to be flush mounted with the pipe wall to prevent them from being severed when rolled onto the spool or from puncturing adjacent hose. (This might be a system similar to the nitrile rubber hose used by NKWD with modified service connections and more substantial trailers.) The savings for such a system would include some manpower that is typically required to install segmented temporary pipeline and the manpower required to chlorinate, flush, and collect and incubate water samples. The benefit to the customer would be two less days temporary pipe was in place by not having a 24-hour contact time with hyper-chlorinated water and a 24-hour incubation period for the water sample.

Major Modification

Utilities that are willing to substantially reassess their pipe lining operations and operating philosophy have the potential to significantly reduce the number of days associated with lining activities. The recommendations below may require close coordination with State and Local officials and the acceptance of risk by the utility for continuous, consistent contractor performance.

Treat Lining Segments Similarly to Main Break Repairs

During a main break repair, utilities generally follow the AWWA Standard for Disinfection Procedures When Cutting Into or Repairing Existing Mains (AWWA C651-99, Sec. 4.7). In general, this involves trench treatment, swabbing with hypochlorite solution, flushing and where practical, slug chlorination. Bacteriological samples are then taken, however, the main is generally returned to service prior to the results of the testing. Thus, the purpose of the bacteriological testing is "to provide a record for determining the procedure's effectiveness" (AWWA C651-99).

The conditions surrounding a main break are much more severe than those encountered during a typical lining operation. Main breaks typically encounter difficult working environments, unknown pipe conditions, and multiple utility components (i.e. waste water, gas, communications, etc.). Lining operations, however, employ a structured schedule on a precisely defined component of the distribution system. Thus, it is likely that a greater opportunity for contamination exists for routine main break repairs than with organized lining activities.

Utilities that apply main break disinfection procedures to newly lined pipe can obtain a substantial time and costs savings. The main break disinfection procedure would eliminate the 24-hour time lag for both the chlorine contact time and the results of the bacteriological tests. This reduced time would be applied to both the lining activity and the temporary pipe setup. As such, if implemented properly, main break disinfection procedures have the potential to eliminate 2-days from the temporary pipe installation task and additional 2-days from the cleaning and lining task. Further analysis and demonstrated procedure effectiveness will be required by State Regulators before these processes are widely accepted and fully implemented.

ULTIMATE GOAL – ELIMINATE TEMPORARY PIPE

By combining a number of the discussed improvements to the cleaning and lining processes the possibility exists to eliminate the need for temporary pipeline from the scope of traditional lining projects. By using a rapid set material such as polyurethane, a coating product can easily be cured the same day it is applied. Combined with rapid chlorination procedures and return to service prior to sample results it is a real possibility to clean, line, and return to service a water pipeline segment in a single workday. Thus, the ultimate modification to the current lining process, which includes the installation of temporary water supply networks, would be to perform the work in such a manner by which temporary pipe is not necessary.

While the methods and materials are in place to eliminate temporary pipe, presumed negative customer response has precluded implementing this procedure. It is unknown if customers would tolerate a single instance of extended water outage (8-hours) in lieu of having temporary pipeline in place for several weeks. Additionally, infrastructure repair could progress at a much faster pace and at a greatly reduced cost to the utility and consumer.

The data collected pertaining to customer complaints suggests that water customers are more tolerant than assumed. This consumer assent is inferred as complaints did not increase with increasing project durations. The number of complaints did, however, rise with the number of breaks in the temporary pipeline. Breaks would almost always be unanticipated and uncontrolled by either the utility or the contractor. Therefore there would likely be no forewarning that water service was about to be interrupted. Unanticipated outages can generate a great number of complaints. Scheduled outages or those carefully planned may generate customer concern, but may not generate the number of complaints as unexpected outages. This data suggests that truthfully informed, well educated customers will tolerate necessary water shut-offs for beneficial construction projects. As such, the concern that customers would not tolerate an 8-hour outage may be unfounded.

Although costs have not been a factor directly considered in this project, costs are indirectly accounted for within manpower estimates. Reducing the man hours to perform specific tasks should have a direct intrinsic effect on overall project cost. As such, eliminating the tasks associated with installing and maintaining temporary pipelines would likely reduce project costs. The pipe lining material that allows a rapid return to service may have a higher initial cost per unit length, but the associated labor savings of not installing temporary pipe and reducing the inherent risks associated with temporary pipe, may offset the increased material costs.

APPENDIX
ASSESSMENT OF DISINFECTION TECHNIQUES

INTRODUCTION

In the course of identifying individual tasks typically associated with potable water pipeline renewal, disinfection was recognized as a task that consumed a good deal of project-time. In most cases, however, the disinfection step did not include a large number of man-hours, but rather time associated with meeting perceived regulatory or operational requirements. Much of the project-time linked to the disinfection process appeared to be a result of 1). implementing an AWWA recommended disinfectant procedure that required an extended contact time (CT) between the pipe's interior wall and a disinfectant, and 2). the time required to incubate each individual water sample as a part of bacteriological testing (usually between 18- and 24-hours). Combined, these two procedures account for approximately 10% of the average project duration on any given pipeline segment. But most importantly, these two procedures, as currently defined, can not be accelerated with additional manpower allocated to these tasks.

While it is essential that newly lined mains meet water quality standards the methods and procedures implemented to meet these standards vary widely among utilities. As an example, disinfection procedural protocol ranges from flushing with distribution system water only (i.e., no additional disinfectant and no contact time), to maintaining an elevated disinfectant residual for a 48-hour contact time. It is noted that utilities in geographic regions with sensitive receiving waters tend to use only protracted flushing as a means to ensure satisfactory water quality following the lining process, while utilities that experience colder water temperatures may implement a 48-hour CT at 50-ppm chlorine.

Recognizing the potential time savings to pipeline renewal operations by simply modifying disinfection procedures, further investigating alternative disinfection methods was warranted. As such, a number of utilities currently engaged in water main renewal were invited to participate in a study designed specifically to assess the effectiveness of alternative disinfection techniques. Listed below is a discussion of the testing program, a listing of the utility participants, the test results and concluding recommendations.

PURPOSE

A number of process modifications identified in the primary document's Conclusions and Recommendations Chapter offered opportunities to reduce the downtime during lining operations. Among those modifications were suggestions to modify the disinfection techniques currently used by many utilities. However, while several utility participants have expressed interest in implementing alternative disinfection methods, they also have concern as to the suitability of the disinfection techniques when applied to newly lined pipe. Thus, the purpose of this Project Continuation Reserve (PCR) was to specifically assess the suitability of two likely alternative disinfection techniques that may be applied to newly lined pipes.

Currently, many utilities disinfect newly lined pipes using a low concentration disinfectant and a 24-hour contact time. However, AWWA Standard for Disinfecting Water Mains (ANSI/AWWA C651-99) indicates that a higher disinfectant concentration and shorter contact time should be equally effective. An acceptable alternative is the Slug Chlorination method (Sec. 4.4.4) that specifies a 100 ppm concentration and a 3-hour contact time. As the disinfection process generally occurs for both the temporary water network and again in the newly lined pipe, any reduction in the time necessary to carry out the disinfection task has the potential to significantly accelerate the overall project schedule.

A second alternative disinfection method presented in the report was to characterize the newly lined mains as "main break repair" (ANSI/AWWA C651-99 Sec. 4.7) and to then follow standard procedures for main break line disinfection. This would involve flushing the newly lined pipe, and where practical, slug chlorination at 300 ppm and 15 minute contact time. At this time, a bacteriological test sample would be obtained and the main would be returned to service. (If the test sample failed, then remedial measures would be enacted). Treating the newly lined main as a "main break repair" would greatly accelerate the main lining process as disinfection would rely primarily on flushing and the main would be returned to service upon identification of a chlorine residual, but prior to results of the bacteriological tests.

Characterizing the newly lined pipe as a "main break repair" and following typical main break disinfection procedures would be a paradigm change for cleaning and lining operations. However, it is believed that a cleaning and lining operation provides less risk for contamination as compared to a main break as newly applied spray linings:

1. Access pits are excavated in a "clean" and organized environment
2. Activities are planned and isolation of the main is carefully executed
3. Generally performed during normal business hours
4. Involve placing "new" material into pipes, not material stored in a yard
5. For some process (like applied linings) pipes are cleaned, reducing potential bacteriological locations within the pipe (tuberculation)

PROCEDURE

To assess alternative disinfection procedures, six utilities actively involved in the pipe lining process were invited to participate in a study. Initially, the study plan was to have utilities collect and analyze bacteriological test samples immediately after flushing and then again after a 1-hour contact time of 100 ppm chlorine concentration. However, the utilities expressed concern that state regulators would not accept the accelerated disinfection methods and mandate a 24-hour contact time. If this were to happen, then utilities would be forced to disinfect their newly lined water mains twice and collect and test as many as 4-water quality samples. Therefore, in order to avoid undue hardship on participating utilities the disinfection process employed for this case study was to flush the newly lined main relying only on the residual chlorine in the drinking water. After flushing, a bacteriological test sample would be obtained. The newly flushed line was then isolated for 24 hours and then a second bacteriological test sample was obtained. It was anticipated that this sample would represent a "worst case" condition and that any subsequent disinfection method would only work to improve water quality. At this point, the utilities were then free to implement any disinfection technique necessary to meet their specific state regulations. In this manner, the cost and time expense to the participating utilities was reduced to two bacteriological test samples and a 24-hour isolation period per lining section monitored. The following is excerpted from the sampling protocol issued to participating utilities:

Sampling Protocol

1) *Clean and line pipe in typical manner*
2) *Fill the newly lined segment with the utility's system water once lining material is sufficiently cured to allow contact with water, using* ***NO*** *additional disinfectant*
3) *Immediately flush newly lined segment until standard chlorine levels are detected at the flushing point and turbidity is within allowable limits*
4) *Obtain water sample from acceptable sampling point*
5) *Close all valves necessary to isolate the segment of main*
6) *Allow newly lined pipe to remain isolated for 24-hours*
7) *Following the 24-hour period open valves and flushing connection, flushing only enough water to insure the sample point is clear of potential contaminates*
8) *Obtain second sample for laboratory testing*
9) *Submit laboratory result data to AwwaRF Study Team*

RESULTS

Six utilities participated in the study to assess the suitability of alternative disinfection techniques for newly lined mains. Each utility tested multiple locations, and in total data was collected from 63 separate lining segments.

Bacteriological Testing

As described, immediately following the lining process participants were requested to thoroughly flush the newly lined pipe to ensure no cleaning debris remained in any outlets or service taps. Water quality samples were then drawn from the main. Chlorine residual testing was performed immediately to ensure the water contained typical levels of disinfectant for the utility. The main was then isolated and left idle for approximately 24-hours. The following day the main was flushed only enough to verify the sample point was free of any contaminants and a second sample was drawn. Again chlorine residual was checked to determine if there had been any significant reduction in available disinfectant, as that could be indicative of bacterial growth in the main. Both the initial and secondary water quality samples were processed by the utility's in-house laboratories following their standard operating procedures.

Table 1 lists the participating utility, the renewal system used by each, the total number of sample locations, and their pass/fail rate. In addition, as a check, results for random distribution system sampling were also included.

As indicated in Table A-1, most of the newly lined pipe segments evaluated passed the bacteriological tests after flushing only. For most utilities, the passing rate for the newly lined section was as high as the passing rate for the routine bacteriological tests of their distribution systems. Any additional disinfectant applied to these newly lined pipe segments should only work to provide a supplement level of care.

Table A.1
Alternative Disinfection Test Results

LINING INFORMATION			BACTERIOLOGICAL TESTS (% Passing)			AVERAGE DISINFECTANT CONCENTRATION (ppm)		
Utility	Lining Material	No. of Sample Locations	After flushing	After 24-hour CT	Random distribution sampling	After flushing	After 24-hour CT	Target Distribution System Residual
Cleveland Division of Water	Cement	25	100	100	98.79	0.94	0.82	0.30
Greensboro Utilities	Epoxy	13	100	100	100.00	0.80	0.26	>0.20
Fayetteville Public Works Commission	Epoxy	3	100	100	100.00	3.72	3.73	2.00
City of Moncton	Polyurethane	6	100	100	99.84	0.74	0.68	0.60
City of St. John	Polyurethane	3	100	100		0.33	0.13	
Northern Kentucky Water District	Epoxy	13	85	62	100.00	0.93	0.18	>0.20
TOTAL		63						

The exception to the high percentage rate of passing bacteriological tests was the data from Northern Kentucky Water District (NKWD). Based on the NKWD data provided, 85 percent of the water quality samples passed the bacteriological tests immediately after lining and flushing, and only 62 percent of the newly lined segments passed the bacteriological tests after the 24 hour isolation period. (It is noted that all newly lined segments subsequently passed bacteriological tests after a 50 ppm concentration and 24-hour contact time.) As the high failure rate in the NKWD data was not observed at other utilities, an investigation into possible failure causes was initiated. Flushing procedures were removing debris and monitoring techniques appeared sound. Thus, no single explanation for the high failure rate has been identified.

Chlorine Demand

In addition to the bacteriological tests samples obtained from the newly lined pipe, concurrent chlorine tests data was also obtained. The purpose of the chlorine testing was first to insure that some disinfectant had entered the newly lined pipe, and secondly, to get an assessment of the chlorine demand during the 24-hour contact time. The disinfectant concentration reported in Table A.1 for both the "After flushing" and "After 24-hour CT" are the average concentration values derived from multiple pipe lining segments.

Based on the data collected, a measurable amount of disinfectant entered the pipes immediately after flushing for all pipe lining segments. For the only cement lining operation in the data set, the disinfectant concentration was reduced by 0.12 ppm during the 24-hour isolation period. For the polyurethane lining operations, disinfectant concentration was reduced by 0.06 and 0.2 ppm for the City of Moncton and the City of St. John, respectively. Two of the three cities using epoxy lining showed a greater chlorine demand during the 24-hour isolation period than other lining systems. For the epoxy lining systems, the change in disinfectant was 0.01 (elevated), 0.54 and 0.75 for Fayetteville, Greensboro and Northern Kentucky, respectively. It is noted that Northern Kentucky Water District's chlorine residual was reduced to nearly undetectable levels during this timeframe.

Utility Data Summary

Six water utility systems provided data representing 63 lining segments (Table A.2). Below is a brief summary of the data results with respect to the individual utility operations.

Cleveland Division of Water supplied the study team with 2-samples from each of 25-selected sample locations within its clean and cement mortar lining project area. Cleveland's standard disinfection routine following their water main lining process mirrors the 'flush-only' protocol of this study. Cleveland, relying on chlorine for disinfection in its distribution system strives to maintain a minimum of 0.3 ppm concentration at the farthest reaches of its pipeline network. Of the 50-total samples collected 100-percent passed Cleveland's bacteriological testing process. Chlorine residual loss during the 24-hour CT averaged 0.19 ppm.

Greensboro Utilities collected a total of 26-samples from 13-sepatate locations within network of water mains included in its clean and epoxy line program. During the bacteriological testing, 100 percent of the samples passed immediately after flushing and 100 percent passed after the 24-hour contact time. During this period, the chlorine demand averaged 0.54 ppm. Greensboro uses free chlorine to disinfect it drinking water supply and maintained an average residual of 0.81 ppm throughout its distribution system during the period samples were collected for this study.

A total of 6-samples included in this study were supplied by Fayetteville Public Works Commission as a part of their 2006 clean and epoxy line project. As with Cleveland and Greensboro, 100-percent of their initial samples and 24-hour follow-up samples passed bacteriological testing. The average reduced chlorine residual concentration in the tested pipeline segments was negligible at 0.00. Fayetteville is the only participant in the study currently using chloramine as a means to maintain a disinfectant residual in their distribution system. The utility maintains a minimum of 2.00 ppm at 95% of its routine sampling stations within their pipeline network.

The City of Moncton is one of the two study participants currently lining cast iron mains with polyurethane. The polyurethane product is similar to epoxy, however, the cure time is a matter of minutes rather than hours. The study includes 12 data points collected form 6 locations within the scope of Moncton's water main renewal project on Pine Street. Moncton's disinfectant goal is to have a concentration of 1.1 ppm chlorine leaving the Treatment Plant, a target of 0.60 ppm in the distribution system with a minimum acceptable level of 0.20 ppm. The reduced chlorine residual concentration in the tested pipeline segments was 0.06 ppm.

The City of St. John, also using polyurethane lining, supplied disinfection data from three lining segments. 100 percent of the samples passed the bacteriological tests at both flushing and after the 24-hour contact time. During the 24-hour isolation period, the chlorine demand averaged 0.2 ppm.

Northern Kentucky Water District provided data from 13 epoxy lining segments. Of these segments, only 85 percent passed bacteriological testing immediately after flushing and only 62 percent passed after the 24-hour isolation period. For routine testing of the distribution system, 100 percent of the samples are reported to pass bacteriological testing. During the 24-hour contact time, the chlorine demand averaged 0.75 ppm, greater than any other utility. Northern Kentucky relies on free chlorine as their disinfectant. The average concentration of chlorine leaving the treatment plant is 1.4 ppm and maintains a residual of greater than 0.20 ppm in the distribution system.

CONCLUSIONS

The purpose of the supplemental testing program was to assess the effectiveness of alternative disinfection procedures applied to newly lined pipe. Alternative disinfection procedures were assessed at six utilities and 63 lining locations. Based on the documented water quality sample results the 'flush only' method may prove to be an acceptable disinfectant procedure following water main lining. With the exception of Northern Kentucky Water District, all of the utilities studied exhibited satisfactory passing bacteriological tests as a result of flushing alone.

As was expected, even newly lined pipes consume some chlorine during a 24-hour idle contact period. Chlorine tests were taken immediately after flushing and then again after the 24-hour idle period. For the one utility using cement lining, the chlorine demand was recorded as 0.12 ppm. For the two utilities, using polyurethane lining, the chlorine demand was recorded as 0.06 and 0.2 ppm. For the three utilities using epoxy lining, the chlorine demand was recorded as 0.01 (elevated), 0.54 and 0.75 ppm.

With the exception of data from Northern Kentucky Water District, the 'flush only' approach resulted in water mains that were satisfactorily disinfected and ready for service. And it appeared that the chlorine demand during the 24-hour isolation period was within a manageable range for most utilities. Thus, it appears reasonable that a "flush only" disinfection procedure may be suitable for some utility systems. However, for other utility systems, a hyper-chlorination combined with a short contact time would provide an additional step towards ensuring sanitary conditions within the newly lined pipes.

It is recognized that the data set used to assess the effectiveness of alternative disinfection procedures was very small and can not represent all conditions. However, the results are encouraging and it is recommended that utilities assess the effectiveness of alternative disinfection procedures within their unique systems.

Table A.2
Utility Disinfection Data

Cleveland

Sample No.	Identification	Pipe Diameter	Pipe Length	Date Sampled	Chlorine PPM	Bacteriological Result	Date Resampled	Chlorine PPM	Bacteriological Result
1	Lakeview Ave. (Detroit/#1664)	6	635	7/6/06	1.14	Pass	7/7/06	0.98	Pass
2	Lakeview Ave. (Riverwood/#1664)	6	481	7/6/06	1.02	Pass	7/7/06	0.87	Pass
3	Fairwood/W. 138	6	900	7/7/06	1.08	Pass	7/10/06	0.92	Pass
4	3724 W. 136	6	1,200	7/7/06	0.89	Pass	7/10/06	0.57	Pass
5	3854 W. 136	6	1,200	7/7/06	1.05	Pass	7/10/06	0.78	Pass
6	3874 W. 136 (Southwood/#3874)	6	900	7/7/06	1.02	Pass	7/10/06	1.20	Pass
7	3874 W. 136 (West Bl./#3874)	6	300	7/7/06	1.02	Pass	7/10/06	0.89	Pass
8	3936 E. 186 N.	8	1,200	7/11/06	0.79	Pass	7/12/06	0.80	Pass
9	3986 E. 188 S.	8	1,200	7/11/06	0.85	Pass	7/12/06	0.82	Pass
10	3896 E. 188 N.	8	1,200	7/25/06	0.91	Pass	7/26/06	0.72	Pass
11	3896 E. 188 S.	8	1,200	7/25/06	0.83	Pass	7/26/06	0.74	Pass
12	3792 E. 188	8	1,200	7/25/06	0.86	Pass	7/26/06	0.73	Pass
13	3900 E. 189	8	1,200	7/25/06	0.86	Pass	7/26/06	0.70	Pass
14	Goodman	6	1,200	8/7/06	0.83	Pass	8/8/06	1.02	Pass
15	Force Ave.	6	800	8/11/06	1.04	Pass	8/14/06	0.73	Pass
16	Ella Ave.	6	600	8/11/06	0.96	Pass	8/14/06	0.90	Pass
17	Gerald	8	295	8/14/06	0.63	Pass	8/15/06	0.66	Pass
18	Maryland Ave.	6	525	8/22/06	0.97	Pass	8/23/06	0.91	Pass
19	Conneticut	6	1,000	8/22/06	1.15	Pass	8/23/06	0.51	Pass
20	Tioga	6	1,000	8/22/06	0.98	Pass	8/23/06	0.97	Pass
21	Turney Rd. (Rosewood/Grandivision)	10	800	8/28/06	0.64	Pass	8/29/06	0.87	Pass
22	Turney Rd. (Rosewood/Macomb)	10	900	8/28/06	0.79	Pass	8/29/06	0.91	Pass
23	Grandivision	6	1,200	8/28/06	0.85	Pass	8/29/06	0.60	Pass
24	Cardwell Ave.	6	1,200	9/13/06	1.21	Pass	9/14/06	0.85	Pass
25	Rosewood	6	1,200	9/13/06	1.02	Pass	9/14/06	0.79	Pass

(continued)

Fayetteville

Sample No.	Identification	Pipe Diameter	Pipe Length	Date Sampled	Chlorine PPM	Bacteriological Result	Date Resampled	Chlorine PPM	Bacteriological Result
1	1306 Ft Bragg Rd.	8	1,100	8/14/06	3.60	Pass	8/15/06	3.80	Pass
2	Westmont	8	1,100	8/14/06	4.05	Pass	8/15/06	4.00	Pass
3	Bethune	8	1,500	8/14/06	3.50	Pass	8/15/06	3.40	Pass

Greensboro

Sample No.	Identification	Pipe Diameter	Pipe Length	Date Sampled	Chlorine PPM	Bacteriological Result	Date Resampled	Chlorine PPM	Bacteriological Result
1	E. Greenway (Chapman/Rolling)	6	480	8/15/06	0.75	Pass	8/16/06	0.10	Pass
2	E. Greenway ((Madison/Rolling)	6	462	8/15/06	0.71	Pass	8/16/06	0.44	Pass
3	Madison (Chapman/E. Greenway)	6	479	8/15/06	0.85	Pass	8/16/06	0.15	Pass
4	Madison (Ridgeway/E. Greenway)	6	677	8/15/06	0.98	Pass	8/16/06	0.20	Pass
5	Market St (Linden/Ridgeway)	8	1,481	8/15/06	0.79	Pass	8/16/06	0.21	Pass
6	Market St (Ridgeway/E. Greenway)	8	455	8/15/06	0.74	Pass	8/16/06	0.47	Pass
7	Rolling Rd (Chapman/Woodbine Ct)	6	295	9/21/06	1.57	Pass	9/22/06	1.12	Pass
8	Madison (Rolling/Madison)	6	445	9/21/06	1.53	Pass	9/22/06	0.79	Pass
9	E. Greenway (Chapman/E. Greenway)	6	535	9/21/06	1.73	Pass	9/22/06	0.38	Pass
10	E. Greenway (Rolling/Friendly)	6	1,093	9/21/06	1.67	Pass	9/22/06	1.64	Pass
11	Woodbine Ct (Rolling/Friendly)	6	540	9/21/06	1.34	Pass	9/22/06	1.67	Pass
12	Tremont Dr (Woodbine/Tremont)	6	359	9/21/06	1.45	Pass	9/22/06	1.84	Pass
13	Market St (E. Greenway/Tremont)	8	1,034	9/21/06	1.49	Pass	9/22/06	1.89	Pass

City of Moncton

Sample No.	Identification	Pipe Diameter	Pipe Length	Date Sampled	Chlorine PPM	Bacteriological Result	Date Resampled	Chlorine PPM	Bacteriological Result
1	40 Pine St.			7/25/06	0.67	Pass	7/26/06	0.59	Pass
2	Pine St. 3			7/26/06	0.99	Pass	7/27/06	0.87	Pass
3	Pine St. 4			7/27/06	0.87	Pass	7/28/06	0.81	Pass
4	Pine St. 5			7/27/06	0.48	Pass	7/28/06	0.81	Pass
5	Pine St. 6			7/28/06	0.73	Fail	7/31/06	0.83	Pass
6	Pine St. 7			7/31/06	0.69	Pass	8/1/06	0.14	Pass

(continued)

St. John

Sample No.	Identification	Pipe Diameter	Pipe Length	Date Sampled	Chlorine PPM	Bacteriological Result	Date Resampled	Chlorine PPM	Bacteriological Result
1	354 Manchester Ave.			5/18/06	0.29	Pass	5/19/06	0.3	Pass
2	15 Carle Ave.			5/18/06	0.2	Pass	5/19/06	0.04	Pass
3	322 Manchester Ave.			5/19/06	0.51	Pass	5/20/06	0.04	Pass

Northern Kentucky Water District

Sample No.	Identification	Pipe Diameter	Pipe Length	Date Sampled	Chlorine PPM	Bacteriological Result	Date Resampled	Chlorine PPM	Bacteriological Result
1	Rossmoyne (Druid/#123)	6	414	5/11/06	0.70	Pass	5/12/06	0.20	Pass
2	Rossmoyne (Windway/#123)	6	1,209	5/11/06	0.80	Pass	5/12/06	0.03	Pass
3	34th & Graff	6	1,230	5/22/06	0.80	Fail	5/23/06	0.20	Fail
4	1035 Emerson	6	1,134	6/19/06	0.22	Fail	6/20/06	0.20	Fail
5	1074 Emerson	6	408	6/19/06	0.70	Pass	6/20/06	0.70	Pass
6	1117 Cleveland	6	1,070	6/27/06	1.30	Pass	6/28/06	0.30	Pass
7	1081 Cleveland	6	760	6/27/06	1.30	Pass	6/28/06	0.05	Pass
8	1048 Rose	6	1,518	7/17/06	1.40	Pass	7/18/06	0.02	Fail
9	312 Eastern	6	475	8/28/06	0.70	Pass	8/29/06	0.30	Pass
10	206 Eastern	6	628	8/28/06	1.10	Pass	8/29/06	0.10	Pass
11	111 Lee	6	722	8/28/06	1.00	Pass	8/29/06	0.10	Pass
12	86 Park	6	350	8/28/06	1.10	Pass	8/29/06	0.00	Fail
13	121 Walnut	6	1,032	8/28/06	1.00	Pass	8/29/06	0.10	Fail

REFERENCES

Allouche, Erez N., K. Bainbridge and I. Moore. 2005. Laboratory Examination of a Cured-In-Place Pressure Pipe Liner for Potable Water Distribution Pipes. In *North American Society for Trenchless Technology, No-Dig Conference & Show, April 24-29 2005, Orlando FL*. Arlington VA. NASTT.

American Water Works Association. 2001. Rehabilitation of Water Mains, Manual of Water Supply Practices: AWWA Manual M28. Denver, Colo.: AWWA.

American Water Works Service Co., Inc. 2002. Deteriorating Buried Infrastructure Management Challenges and Strategies.

AWWA (American Water Works Association. 1999. AWWA Standard for Disinfecting Water Mains.ANSI/AWWA C651-99. Denver, CO: AWWA.

Boxall, J.B., A. Saul and P. Skipworth. 2004. Modeling for Hydraulic Capacity. *Journal AWWA*. 94(4):161-169

Bueno, Sharon M.. 2005. Jacksonville's Sewer Problem. *Trenchless Technology Special Supplement*. November:8-10

Calvin, Todd. 2004. Teamwork Solves Sinkhole Crisis, *Underground Construction*. May:57-60.

Carpenter, Carol. 2004. Christmas Day Fire: Utility Learns Lesson of the Valves. *American Water Works Association, Opflow*. 30(4):16-17.

Conroy, Paul J., D. Hughes, and I. Wilson, 1995. Demonstration of an Innovative Water Main Rehab Technique: Epoxy Lining. Denver, Colo.: AwwaRF and AWWA.

Cote, Benoit. 2005. Water Main Rehabilitation Using CIPP Structural Lining Technology. In *North American Society for Trenchless Technology, No-Dig Conference & Show, April 24-29 2005, Orlando FL*. Arlington VA. NASTT.

Crabtree, Douge, 2005. Rehabilitation of a 16" Water Pipeline Utilizing the U-Liner High Density Polyethelene Liner System. In *North American Society for Trenchless Technology, No-Dig Conference & Show, April 24-29 2005, Orlando FL*. Arlington VA. NASTT.

Deb, K., Arun. 2001. New Techniques for Precisely Locating Buried Infrastructure. Denver, Colo.: AwwaRF and AWWA.

Deb, Arun K., F. Grablutz, Y. Hasit, and J Snyder. 2002. Prioritizing Water Main Replacement and Rehabilitation. Denver, Colo.: AwwaRF and AWWA.

Deb, K., Arun, Y. Hasit, C. Norris. 1999. Demonstration of an Innovative Water Main Renewal Technique. Denver, Colo.: AwwaRF and AWWA.

Deb, K., Arun, Y. Hasit, H. Schoser and J. Snyder. 2002. Decision Support System for Distribution System Piping Renewal. Denver, Colo.: AwwaRF and AWWA.

Deb, K., Arun, R. Herz, Y. Hasit, and F. Grablutz. 1998. Quantifying Future Rehabilitation and Replacement Need of Water Mains. Denver, Colo.: AwwaRF and AWWA.

Deb, K., Arun, K. Momberger, Y. Hasit, and F. Grablutz. 2000. Guidance for Management of Distribution Operation and Maintenance. Denver, Colo.: AwwaRF and AWWA.

Deb, K., Arun, J. Snyder, and J. Chelius. 1990. Assessment of Existing and Developing Water Main Rehab Practices. Denver, Colo.: AwwaRF and AWWA.

EPA. 1999. Understanding the Safe Drinking Water Act. [Online]. Available: < http://www.epa.gov/safewater/sdwa/30th/factsheets/understand.html>. [cited June 8, 2005]

EPA White Paper. 2005. New or Repaired Water Mains. [Online]. Available: < http://www.epa.gov/ogwdw/tcr/pdf/maincontam.pdf>. [cited June 8, 2005]

Evans, Gerry and B. Wiens. 2005. RAAM-Remote Anode Attachment Method. In *North American Society for Trenchless Technology, No-Dig Conference & Show, April 24-29 2005, Orlando FL.* Arlington VA. NASTT.

Frey Water Engineering, Inc. 2000. Optimizing a Main Replacement. The Optimatics Letter. Arlington Heights, IL. 10:1-3.

Griffin, Jeff. 2002. Pipebursting Installations with Prechlorinated HDPE, *Underground Construction: November.*

Grigg, Neil S. 2005. Assessment and Renewal of Water Distribution Systems. *Journal AWWA.* 97(2):58-67

Grob, Robert. 2004. Air Scouring-An Alternative for Water Main Cleaning. *WaterWorld*, 36, November.

Kampbell, Ed. 2005. Cured in Place Pipe Designed to Last. In *North American Society for Trenchless Technology, No-Dig Conference & Show, April 24-29 2005, Orlando FL.* Arlington VA. NASTT.

Kennedy, Robert 2005. Fire Ratings – What is their Purpose. *Louisville Water Company, Newsleaks.* Louisville, Kentucky. January 24:1.

Khan, Salam A. 2005. Trouble Shooting for Liner Installation: Observations and Recommendations Based on Field Experiences. In *North American Society for Trenchless Technology, No-Dig Conference & Show, April 24-29 2005, Orlando FL.* Arlington VA. NASTT.

Kirmeyer, Gregory J., M. Friedman, and J. Clement. 2000. Guidance Manual for Maintaining Distribution System Water Quality. Denver, Colo.: AwwaRF and AWWA.

Kosub, Kavid. 2003. Canadian Rockies: Preparing for the Unexpected. *Underground Construction.* November:27-30.

Kwan, Albert, and K. Chua. 2005. Pipe Bursting for Pipe Rehabilitation – A Desktop Study on Cost Effectiveness. In *North American Society for Trenchless Technology, No-Dig Conference & Show, April 24-29 2005, Orlando FL.* Arlington VA. NASTT.

Methods, Haestad, T. Walski, D. Chase, D. Savic, W. Grayman, S. Beckwith and E. Koelle. 2003. *Advanced Water Distribution Modeling and Management*, Waterbury, CT: Haestead Press.

Rajani, Balvant, and Y. Kleiner. 2004. Alternative Strategies for Pipeline Maintenance/Renewal. In *American Water Works Association 2004 Annual conference, June 13-17, Orlando, FL.* Denver Colo. AWWA.

Rush, James, W. 2005. Development of a Performance Specification in Amarillo, Texas. *Trenchless Technology.* March. 25-26.

Rush, James, W. 2003. Trenchless In Drinking Water Rehabilitation: An Idea Whose Time Has Come? Trenchless Technology. May. 57-58.

Saber, Azis and R. Sterling. 2005. Monitor Pipe Liner Performance Using Fiber Optic Sensors. In *North American Society for Trenchless Technology, No-Dig Conference & Show, April 24-29 2005, Orlando* FL. Arlington VA. NASTT.

Saegrov, Sveinung. 2004. Computer Aided Rehabilitation of Water Networks. In *American Water Works Association 2004 Annual conference, June 13-17, Orlando, FL.* Denver Colo. AWWA.

Salvo, Piero. 2005. Watermain Rehabilitation – Evolving Project Costs, Where Does the Buck Stop. In *North American Society for Trenchless Technology, No-Dig Conference & Show, April 24-29 2005, Orlando FL.* Arlington VA. NASTT.

Selvakumar, Ariamalar, R. Clark and M. Sivaganesan. 2002. Costs for Water Supply Distribution System Rehabilitation [Online]. Available: < http://www.epa.gov/ORD/NRMRL/pubs/600ja02406/600ja02406.pdf>. [cited June 8, 2005].

Simicevic, Jadranka and R. Sterling. 2001. *Guidelines for Pipe Bursting.* Technical Report #2001.02. U.S. Army Corps of Engineers, Engineering Research and Development Center. [Online]. Available: http://www.latech.edu/tech/engr/ttc/publications/guidelines_pb_im_pr/bursting.pdf. [cited June 8, 2005].

Spelay, Karen and Sharp, J. 2004. Breaking the Ice: Deep Water, Low Temperatures Hamper Main Break Repair. *American Water Works Association, Opflow.* 30(7):1-6.

Stubbart, John. 2004. How Should We Size Service Lines? *American Water Works Association, Opflow.* 30(4):8-9.

Trenchless Technology Special Report. 2005. Jacksonville's JEA – Jacksonville's Approach to Solving Its Infrastructure Crisis. November.

Tokuno, Shin-ichi. 2003. Chlorine Residual Boosting in Distribution Water: Problems with Chlorine Application and Disinfection Byproducts - Part 1. *Water Engineering & Management.* April:25-29.

Tokuno, Shin-ichi. 2003. Chlorine Residual Boosting in Distribution Water: Problems with Chlorine Application and Disinfection Byproducts - Part 2 [Online]. http://www.waterinfocenter.com/wwd/index.cfm/powergrid/rfah=|cfap=/fuseaction/showArticle/articleID/4062. [cited June 8, 2005].

Twort, A.C., F. Law, F. Crowley, and D. Ratnayaka. 1997. *Water Supply, 4th Edition.* New York: John Wiley & Sons, Inc.

U. S. Department of Health and Human Services. 1997. Food Code. [Online]. Available: http://vm.cfsan.fda.gov/~dms/fc-toc.html. [cited June 8, 2005].

Utility Maintenance. 2002. Hydrant Water. [Online]. Available: < http://www.utilitymaintenance.com/hydrant_water.html>. [cited June 8, 2005].

Wage, Jeff, A. Mayer, and D. Mellert. 2005. Installing Pre-Chlorinated Water Mains With Static Pull Pipe Bursting Systems in Fort Pierce, Florida. In *North American Society for Trenchless Technology, No-Dig Conference & Show, April 24-29 2005, Orlando FL.* Arlington VA. NASTT.

Warren Associates (Pipelines) Limited. 2000. *In Situ Rapid-Setting Polymeric Lining, Operational guidelines and Code of Practice.* Devizes, Wiltshire, UK: Warren Associates (Pipelines) Ltd.

Zubko, Nick. 2005. CemPipe Ltd. - Creating a Middle Ground in Waterline Rehabilitation. *Trenchless Technology International.* January:10-13

Zubko, Nick. 2004. Sewer Pipe Usage Survey. Trenchless Technology. August.:20-28.

ABBREVIATIONS

AACDPW	Anne Arundel County, Maryland, Bureau of Public Works
ADA	American Disabilities Act
ANSI	American National Standards Institute
AwwaRF	Awwa Research Foundation
AWWA	American Water Works Association
CCTV	Closed Circuit Television
CIPP	Cast In Place Pipe
CMU	Charlotte-Mecklenburg Utilities
CT	Contact Time
FAQ	Frequently Asked Questions
FDA	Food and Drug Administration
FDEP	Florida Department of Environmental Protection
ft	feet
GCWW	Greater Cincinnati Water Works
GIS	Geographical Information System
HDPE	High Density Polyethylene
IQR	Inter Quartile Range
KY	Kentucky
L	Liter
LWC	Louisville Water Company
MD	Maryland
mg	milligrams
CMU	Omaha, Nebraska Municipal Utilities District
MWS	Nashville, Tennessee Metro Water Services
NC	North Carolina
NE	Nebraska
NKWD	Northern Kentucky Water District
NSF	NSF International
OH	Ohio
PCCP	Pre Cast Concrete Pipe

ppm	Parts Per Million
PRV	Pressure Reducing Valves
PU	Polyurethane
PVC	Polyvinyl Chloride
SDWA	Safe Drinking Water Act
sec	second
SER	Systems Evaluation and Rehabilitation
TN	Tennessee
WIR	Water Infrastructure Rehabilitation Program

IWA Publishing's authorised EU representative for General Product Safety Regulations is Diane D'Arras, 15 rue Duret, 75116 Paris, France, e-mail: safety@iwap.co.uk.

Printed and bound by CPI Group (UK) Ltd, Croydon, CR0 4YY
19/09/2026
02219351-0003